ANIMAL SOCIETIES AND EVOLUTION

Readings from
**SCIENTIFIC
AMERICAN**

ANIMAL SOCIETIES
AND EVOLUTION

With an Introduction by
Howard Topoff
Hunter College of the City University of New York
American Museum of Natural History

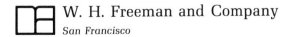 W. H. Freeman and Company
San Francisco

Most of the SCIENTIFIC AMERICAN articles in *Animal Societies and Evolution* are available as separate Offprints. For a complete list of articles now available as Offprints, write to W. H. Freeman and Company, 660 Market Street, San Francisco, California 94104.

Library of Congress Cataloging in Publication Data
Main entry under title:

Animal societies and evolution.

    Bibliography: p.
    Includes index.
    1. Animal societies—Addresses, essays, lectures.
2. Evolution—Addresses, essays, lectures. 3. Social
behavior in animals—Addresses, essays, lectures.
I. Topoff, Howard R. II. Scientific American.
QL775.A54   591.52'4   81-7830
ISBN 0-7167-1333-0   AACR2
ISBN 0-7167-1334-9 (pbk.)

Printed in the United States of America

9 8 7 6 5 4 3 2 1

CONTENTS

Note on cross-references to SCIENTIFIC AMERICAN *articles:* Articles included in this book are referred to by title and page number; articles not included in this book but available as Offprints are referred to by title and offprint number; articles not included in this book and not available as Offprints are referred to by title and date of publication.

PREFACE

What do human beings, ants, and slime molds have in common? Despite their vast differences in structure, physiology, and ecology, all three consist of individuals whose behavior is sufficiently coordinated for the group to be called a society. Elucidating the processes by which this coordination is achieved forms the basis for the comparative study of social behavior, and therefore for this collection of articles from *Scientific American*.

Surprisingly, one of the most important insights into the nature of social organization stems from observations on the behavior of insects, conducted during the first decade of this century by the French biologist Émile Roubaud. When a larva of the wasp *Belanogaster* is fed by an adult, the larva secretes in return a droplet of salivary fluid that the adult avidly ingests. Roubaud concluded that the cooperative rearing of larvae, as well as other associations between the adults and the brood, could be explained by the "attraction" of the wasps to the larval secretions. Several years later, William Morton Wheeler proposed the term "trophallaxis" for the mutual exchange of food that was known to occur in many other social insects. In subsequent years, the concept of trophallaxis was greatly extended. At first, Wheeler himself added the mutual exchange of chemical secretions that were not necessarily involved in feeding. Still later, the comparative psychologist T. C. Schneirla recognized that group members could also achieve behavioral integration by the mutual exchange of tactile, visual, and auditory stimuli. As a result, Schneirla proposed the more general term "reciprocal stimulation," and noted that Wheeler's trophallaxis was only one of many patterns of communication that all social animal species utilize to achieve group integration. Although these concepts may seem a bit superfluous by present-day standards, their application earlier in this century clearly helped eliminate vague and subjective ideas such as "group instinct" as explanations for social behavior. By the shift in emphasis to specific processes of communication that result in the formation of social bonds, the path was clear for an objective, comparative approach to the study of social behavior in diverse animal species.

The articles selected for this book reflect a rather broad definition of social behavior. Because it includes theoretical issues, as well as empirical studies on a variety of invertebrate and vertebrate species, the book can serve as an excellent source of supplementary reading for courses in animal behavior, comparative psychology, and anthropology. On a more general level, however, the book should also be useful to the nonprofessional person, whose primary interest may be a better understanding of the similarities and differences between humans and other social species. I hope the articles will illustrate that an objective approach to the study of behavior can still be accompanied by an equally subjective fascination and excitement with research in this area. After all, we are the only species of animal that has evolved the capability of contemplating other social species. Let's make the most of it!

June 1981   Howard Topoff

ANIMAL SOCIETIES AND EVOLUTION

INTRODUCTION

Introducing a book about the social behavior of animals should be easy. After all, our own existence as members of a complex social species undoubtedly enhances our appreciation of the characteristics and benefits of social behavior, wherever it occurs in the animal world. But what is meant by social behavior? The category "social" is an arbitrary concept that includes behavioral patterns of enormous diversity. The most traditional view of social behavior is based on a rather narrow definition, generally including only those animals that live in permanent groups. Regardless of whether the species are colonies of social insects, schools of fish, or troops of baboons, coordinated behavior clearly facilitates food acquisition, defense, and mate selection, as well as many other activities necessary for individual and group survival. But along with the benefits of group living come some disadvantages such as increased competition for food, mates, and other resources, all of which can lead to severe in-fighting among conspecifics. Indeed, one of the principal goals of research on social behavior is to elucidate the behavioral processes used by organisms to balance the advantages and disadvantages of communal living.

In addition to permanent group living, social behavior has evolved in many other forms. If we use a broad definition, social behavior actually becomes the rule in nature. After all, most organisms interact at some stage of their life cycles with other species members. Often, these social interactions are fleeting, as when an otherwise solitary male and female butterfly come together during a brief period each year for mating. On the other hand, if we consider parent–offspring relationships, social interactions are often quite prolonged. Consider, for example, the process in precocial birds called imprinting, where a newly hatched individual forms a strong attachment to its female parent and follows closely behind her for many weeks. The role of development in the formation of social bonds between parents and offspring is also important in many insect and mammalian species.

But even the broadest definition of social behavior has its boundaries. Animals may often be in groups, but such groups do not always signify the establishment of social bonds. For example, a group of fish thrown together by wave action into a common tide pool is a passive aggregation, not a social unit. Similarly, the term social is not appropriate for active aggregations, such as the many moths that are lured to a porch light on a summer night. Each moth is responding independently to an external stimulus, namely the light, not to other moths. True social behavior, regardless of its duration or level of complexity, is based on communication among individuals of the species. The arbitrariness of the term social is also evident by the practice of restricting its use to behavioral interactions within a species. Obviously, this eliminates interactions between predator and prey, parasite and host, and numerous other interspecific animal relationships. Finally, we also recognize that the very distinction between socially directed behavior and patterns concerned more with individual maintenance and survival is frequently blurred. Thus, an organism that is feeding, running from a predator, or even urinating may simultaneously be conveying information that can affect the behavior of other individuals.

The study of social behavior has recently shifted emphasis by including the concepts and methodology that are subsumed under the heading "socio-biology." And like any other radical change in science, it has generated its fair share of both excitement and controversy. The principal goal of sociobiology is straightforward: to systematically study the biological basis of all social behavior by the application of principles derived from population genetics and evolutionary theory. By emphasizing the role of convergent evolution, some sociobiologists stress the functional similarities in social behavior among diverse animal species.

To illustrate this approach, let's compare army ants and human beings, two species in which most individuals spend their entire lives within the same social group. During the course of evolution, both species have indepen-dently acquired analogous adaptations for group living. These include sufficient longevity to enable social bonds to be formed among successive generations, and a division of labor based on individuals that are specialized for such activities as food gathering, group defense, and care of young. Clearly, the success of these social adaptations depends on precise commu-nicative integration among all individuals. In army ants, as in many other social insects, pheromones (chemicals secreted by one individual that affect the behavior of conspecifics) and tactile stimuli are the primary avenues for social communication. Visual and auditory stimuli, by contrast, play virtually no role in the behavioral integration of army ants. In humans, as we are well aware, the emphasis is reversed. Obviously, we too utilize olfaction and touch for social encounters. But it is the combination of a visual system and a vocal language that forms the cornerstone of human social behavior.

The limitations of the analogy between the social behavior of ants and humans are actually far more extensive than the mere utilization of con-trasting sensory processes. The nervous system of ants and humans is so different that they possess qualitatively different capacities for behavioral integration and plasticity. The division of labor in army ants is based on the development of discrete morphological castes. In each colony, for instance, there is typically one large, fertile female and thousands of smaller workers that are sterile. In some species, the largest of these workers have huge heads and sickle-shaped jaws. They are popularly known as "soldiers," and serve as the colony's first line of defense. Intermediate-sized army ants are most nu-merous and comprise most of the group taking part in the predatory raids for food. Finally, the smallest workers spend most of their time tending the brood inside the nest. In humans, by contrast, childbirth and nursing are probably the only social roles that are based on specific biological character-istics. Virtually every other function in human society depends more on cul-tural factors than on specialized morphological and physiological adaptations.

The conclusion drawn from this example is a reminder that the evo-lutionary process selects for adaptive outcomes. The particular mechanisms and processes utilized to achieve those outcomes may be qualitatively differ-ent for species representing contrasting phyletic levels. A truly comparative science of social behavior should not emphasize analogous functional simi-larities at the expense of elucidating species' differences in mechanisms subserving social adaptations. A comparative approach that balances simi-larities in evolutionary outcomes with differences in proximate mechanisms will undoubtedly yield the most comprehensive understanding of social be-havior in the many diverse animal species.

A second area of interest to sociobiologists involves the genetic basis of social behavior. Its origin is traced directly to Charles Darwin's difficulty in accounting for the evolution of worker sterility in social insects. Since natural selection is based on the differential reproduction of individuals, it was obviously difficult to imagine how a trait conferring zero reproductive potential could persist beyond the generation in which it first appeared. Darwin solved the problem simply by shifting the level of action of natural selection from the individual workers to the colony as a whole.

In 1964, the population geneticist W. D. Hamilton proposed an alternative explanation, based on a new concept called kin selection. The essence of kin selection is that relatives share a certain proportion of their genes. In humans, for example, parents share 50 percent of their genes with all offspring. The more distant relationship, the lower the proportion of shared genes. Hamilton bridged the gap between the evolution of sterile castes and kin selection by pointing out the genetic consequences of the peculiar mode of sex determination in the insect order Hymenoptera (which includes ants, bees, and wasps). Female Hymenoptera are diploid, having developed from fertilized eggs. Males, by contrast, develop from unfertilized eggs and are therefore haploid. If a queen mates with only one male during her life, all the sperm she receives will be genetically identical. As a result, the female offspring of a hymenopteran queen will share an average of 75 percent of their genes with each other. The queen and her offspring, by comparison, share only 50 percent of their genes. Therefore, those female offspring may actually increase their inclusive fitness more by helping to raise additional sisters (some of whom will become queens) than by an equal amount of care given to their own offspring. In other words, if a gene that promotes sterility arises in a population, it can be maintained by natural selection if the sterile individuals' behavior increases the reproductive output of close relatives.

Since the publication of *The Origin of Species*, animal behaviorists have recognized the similarity between Darwin's problem with sterile insect castes and many other patterns of social behavior. After all, the very term social connotes cooperation, sharing, helping, and even sacrifice, all of which may contribute somewhat to the reproductive fitness of another individual. Indeed, seen in this light, the behavior of sterile insects is merely one end of a continuum of behaviors that sociobiologists call "altruistic." Of the several genetic theories that have been postulated to account for altruistic behavior, we have already examined kin selection. A second is called "reciprocal altruism," and it is really a variation of the more traditional individual selection model. In reciprocal altruism, an individual that performs a helpful act at reproductive cost to itself will be compensated later by the performance of a similar act by the genetically unrelated individual. Thus, if a bird emitted an alarm call after seeing a predator, and thereby saved the life of another group member, its reduction in individual fitness would be temporary. Upon subsequent predatory attacks, the individual saved in the first encounter might emit the alarm call.

Although it can be genuinely useful to view social behavior from an evolutionary perspective, we must guard against the exclusive use of sociobiological principles to elucidate social processes. In the first place, kin selection, reciprocal altruism, and indeed all other genetic models are concerned only with the maintenance of social behavior in a population through the operation of natural selection. But before selection can act, the species must possess numerous structural, physiological, and behavioral adaptations that promote communication and the formation of social bonds. That natural selection cannot "guarantee" the development of these social adaptations is best illustrated by noting that the vast majority of Hymenoptera are solitary species, despite the presence of the same haplodiploid mechanism of sex determination found in social forms.

Another problem, even broader in scope, stems from the tendency of biologists to view evolution from the limited perspective of changes in a population's gene pool. An unfortunate consequence of this approach is that the organism tends to be ignored, or at best is relegated to little more than a physical protector of the genotype. Undoubtedly, this is what accounts for the rather one-sided approach to behavior, in which the same genetic models are used to explain the evolution of analogous patterns of social behavior in animals so distantly related as ants and baboons. An additional consequence of what evolutionary biologists call "gene thinking" is that behavior is regarded as fundamentally identical to many morphological adaptations. By this I mean that behavior, even complex social activities, is treated as an intrinsic characteristic of organisms, tightly regulated by the individual's genotype. To be sure, these assumptions have not gone unchallenged, especially when attempts are made to incorporate human behavior into a sociobiological framework. But why the objection? Are we simply afraid to admit that human beings are part of the natural world? Not at all! There is no denying that humans look different and behave differently from all other species, and that part of this disparity is due to genetic differences. But it is simply incorrect to turn the story around and proclaim that each pattern of human behavior, such as warfare, parental roles, sexual preferences, and even birth control, has been individually selected for and is therefore coded by a specific region of the genome. The fact is that humans, like many other vertebrate species, have long life cycles and have not been subjected to the rigorous breeding studies that characterize research with fruit flies, mice, and many other familiar laboratory-raised animals. As a result, we have practically no information concerning the relationship between genes and normal patterns of social behavior. In the absence of such information, it is not appropriate to assume a genetic basis for behavior simply because it is adaptive or because it fits one or more sociobiological models. After all, no one would deny the adaptive value of a sexually mature bird emitting its species-typical vocalization during courtship. Nevertheless, it is well known that in many species the song is individually acquired by imitation during the development of the young bird. As an example a bit closer to home, just imagine all the relationships we routinely have with other people that could easily fit into the sociobiological category of reciprocal altruism. These activities might include calling the police after seeing a robber grab a person's wallet or even lending money to a friend for use as a down payment on a house. But instead of invoking a specific gene complex for each of these activities, why not explain them instead as a prime example of cultural learning? The human behavioral repertoire is replete with social interactions that are adaptive and that undoubtedly reflect the outcome of evolutionary processes. But instead of hypothesizing the modification of specific genes for aggression, mate selection, territoriality, dominance hierarchies, and many other behaviors, the real "target" of natural selection may very well have been our large brain and the resulting potential for cultural learning and behavioral flexibility. From this perspective, we would be no more justified in proposing the existence of genes for warlike aggression than we would in proposing the existence of genes for peacemaking. Instead of conceptualizing culture as the antithesis of biology, our extraordinary capacity for both behavioral and environmental modification should be seen as an integral part of our biology and therefore the real outcome of our evolutionary history.

Our ultimate objectives for understanding social behavior do not necessarily change when our interest shifts from humans to other animal species. But if we allow behavior to get lost in the genetic shuffle and attempt to explain all behavior in terms of similar evolutionary concepts, we run the risk of constructing a large theoretical structure before we have a firm factual foundation. It might very well be that techniques and principles applicable to one group of animals might not be the best conceptual tools for elucidating the behavior of groups representing different levels of evolutionary history. Comparative social behavior is a relatively young discipline. It is only through an approach in which studies of proximate behavioral processes complement evolutionary principles that we will ultimately comprehend the relative similarity and uniqueness of all animal species. This is the approach that I hope will be communicated by the following series of articles, reprinted from *Scientific American*.

Because communication among species is the raw material for all forms of social behavior, we begin with Edward O. Wilson's introduction to "Animal Communication." Wilson's approach is novel and effective. He starts by comparing human language with one of the most sophisticated communication systems found in the animal world: the waggle dance of the honeybee. Although some superficial similarities are noted, like the ability to communicate about objects removed in time and space, the waggle dance is severely limited in comparison with human verbal language. Indeed, the human potential for combining a relatively small number of sounds to communicate a virtually infinite variety of ideas is unparalleled in any other species.

The second article, "The Evolution of Behavior," by John Maynard Smith, explores in detail the genetic basis for kin selection and other evolutionary routes to social behavior. Although somewhat technical, the article lucidly provides the reader with the conceptual tools for understanding much of the current literature in sociobiology.

Leaving theory behind, we next move to a series of articles dealing with empirical studies of social behavior in specific animal groups. Thus, the third article explores the fascinating world of "Social Spiders." As J. Wesley Burgess points out, no known species of spider has achieved the degree of sociality common to ants, termites, and many species of bees and wasps. Nevertheless, in a few spiders, such as the African species *Agelena consociata*, groups of individuals construct a communal web and readily cooperate in prey capture and feeding. Although no permanent caste system is present, an individual spider's role in these group processes does change with age. Through studies of such communal spiders, Burgess provides valuable insight concerning the evolutionary stages that other arthropods may have encountered on their road to complete sociality.

In the next two articles, we follow the biblical advice of Solomon and "go to the ant." To many nonscientists, ants are still thought of as coming in only two varieties—ferocious red and harmless black. In reality, there are presently more than 8,000 species of ants in the world, and their geographical distribution is exceeded only by that of humans. More importantly, ants exhibit such diverse life-styles and social behavior that few generalizations are possible even within this single insect family. Army ants, for example, are characterized by group predation and periodic emigrations to new nesting sites. The comparative psychologist T. C. Schneirla (1902–1968) conducted the first rigorous field studies of tropical army ants and found that their nomadic behavior was influenced by chemical stimulation from the colony's immense brood. In "The Social Behavior of Army Ants," Howard Topoff extends the work of Schneirla through studies of division of labor and mass chemical communication. In "Weaver Ants," by Berthold K. Hölldobler and Edward O. Wilson, we are introduced to a very different kind of ant social organization, even though the use of pheromones for behavioral recruitment is no less developed. In weaver ants too, there is a close interdependence between adult workers and the larval brood. Perhaps the most outstanding expression of this relationship occurs when the mature workers use their own larvae as "tools" for joining the edges of leaves and making compartments in which the colony lives.

Beginning with "The Schooling of Fishes," our attention turns to the social behavior of vertebrates. The most conspicuous features of schooling fish are their constancy in orientation and their synchronization of speed and direction of movement. In this article, Evelyn Shaw describes experiments concerning the role of visual communication in keeping the school together as a coordinated unit. By comparing the behavior of immature fish that were reared in physical and visual isolation with that of socially reared individuals, Shaw vividly illustrates how developmental studies of communication can clarify the nature of social bonds in adult organisms.

The next article, "'Imprinting' in a Natural Laboratory," by Eckhard H. Hess, continues the theme of ontogenetic analysis, but in doing so, we temporarily switch gears. Instead of a focus on social groups consisting predominantly of adult organisms, the emphasis here is on parent–offspring interaction in birds. Almost all modern research on imprinting stems from the observations of Konrad Lorenz on the behavior of newly hatched goslings. Since the 1930s, Lorenz' conclusion about the permanence of imprinting has been disputed. Indeed, many laboratory studies of the phenomenon have not demonstrated irreversible and exclusive attachment to the object utilized as an artificial parent. According to Hess, these conflicting results can be attributed to the "unnatural" laboratory setting in which most studies have been conducted, and his solution is to take the laboratory into the field. In the natural environment, Hess found that the early social experience of ducklings with their female parent were not alterable by subsequent human intervention. Another aspect of Hess' research again highlights the value of a developmental approach to social behavior. It has been known for some time that a clutch of mallard eggs reared in the laboratory hatches over a two- or three-day period. In the natural environment, all the eggs in a clutch hatch at approximately the same time. It appears that synchronization of hatching is achieved through a subtle form of vocal communication between the mother and the unhatched ducklings. Moreover, this prehatching vocal interaction may also facilitate the subsequent development of visual communication and the eventual imprinting of the offspring to the female parent.

Although the approach to social behavior in most of the articles in this volume has been through studies of communication among the individuals of the group, social interactions can also be influenced by the physical and biotic environment that the group encounters. In "The Social Ecology of Coyotes," Marc Bekoff and Michael C. Wells take us on a trip to Blacktail Butte in Grand Teton National Park. Their field research shows that coyotes exhibit a highly variable pattern of social organization, ranging from solitary existence to relatively stable groups. The particular mode depends on the season and the food supply. Social interactions are limited during the summer, as coyotes hunt and kill small prey distributed widely over a large area. In winter, as the food shifts to large, dead prey (like elk) found as isolated clumps, the coyotes form into groups containing up to seven individuals.

In the next article, "Dolphins," we return to the sea. For many years, observations on dolphin social behavior were based on animals in captivity. As you can imagine, it is extremely difficult to accumulate data on the social system of a species that communicates primarily underwater. In his field studies of the Atlantic bottlenose dolphin, Bernd Würsig relied heavily on underwater recordings of dolphin vocalizations. To recognize individual dolphins, photographs were taken of their dorsal fins, which protrude above the surface of the water. The results of this study showed that dolphins utilize complex signals in a rich behavioral repertoire. For example, although dolphins, like many other vertebrate species, exhibit a dominance hierarchy, it is not necessarily related to reproductive behavior. Instead, dominance behavior enables certain individuals to organize other members for group protection from predators. The dolphin society is an open one, with individuals frequently moving among groups. When we add a polygamous mating system to the dolphins' group dynamics, their social system exhibits marked parallels to that found in African chimpanzees and many other terrestrial mammals.

In the last article, we turn to the primates, the closest living relatives of human beings. In their studies of "The Social Life of Baboons," S. L. Washburn and Irven DeVore observed 30 troops in the Amboseli game reserve in Kenya. Not surprisingly, social behavior within each troop turns out to be the principal adaptation for individual survival. Here is another instance in which the group affords protection from predators and an increased knowledge of the territory it occupies. Each day the troop may cover up to four miles, roaming from rest areas to feeding grounds. The group moves quickly and does not wait for sick or injured individuals. Once separated from the group, the chances of death increase substantially. In addition to examining the ecology, social system, and communication behavior of the baboons, the article concludes with a comparison of these features with human society.

Animal Communication

by Edward O. Wilson
September 1972

*Animals ranging from insects to mammals
communicate by means of chemicals, movements, and
sounds. Man also uses these modes of communication,
but he adds his own unique kind of language.*

The most instructive way to view the communication systems of animals is to compare these systems first with human language. With our own unique verbal system as a standard of reference we can define the limits of animal communication in terms of the properties it rarely—or never—displays. Consider the way I address you now. Each word I use has been assigned a specific meaning by a particular culture and transmitted to us down through generations by learning. What is truly unique is the very large number of such words and the potential for creating new ones to denote any number of additional objects and concepts. This potential is quite literally infinite. To take an example from mathematics, we can coin a nonsense word for any number we choose (as in the case of the googol, which designates a 1 followed by 100 zeros). Human beings utter their words sequentially in phrases and sentences that generate, according to complex rules also determined at least partly by the culture, a vastly larger array of messages than is provided by the mere summed meanings of the words themselves. With these messages it is possible to talk about the language itself, an achievement we are utilizing here. It is also possible to project an endless number of unreal images: fiction or lies, speculation or fraud, idealism or demagogy, the definition depending on whether or not the communicator informs the listener of his intention to speak falsely.

Now contrast this with one of the most sophisticated of all animal communication systems, the celebrated waggle dance of the honeybee (*Apis mellifera*), first decoded in 1945 by the German biologist Karl von Frisch. When a foraging worker bee returns from the field after discovering a food source (or, in the course of swarming, a desirable new nest

site) at some distance from the hive, she indicates the location of this target to her fellow workers by performing the waggle dance. The pattern of her movement is a figure eight repeated over and over again in the midst of crowds of sister workers. The most distinctive and informative element of the dance is the straight run (the middle of the figure eight), which is given a particular emphasis by a rapid lateral vibration of the body (the waggle) that is greatest at the tip of the abdomen and least marked at the head.

The complete back-and-forth shake of the body is performed 13 to 15 times per second. At the same time the bee emits an audible buzzing sound by vibrating its wings. The straight run represents, quite simply, a miniaturized version of the flight from the hive to the target. It points directly at the target if the bee is dancing outside the hive on a horizontal surface. (The position of the sun with respect to the straight run provides the required orientation.) If the bee is on a vertical surface inside the darkened hive, the straight run points at the appropriate angle away from the vertical, so that gravity temporarily replaces the sun as the orientation cue.

The straight run also provides information on the distance of the target from the hive, by means of the following additional parameter: the farther away the goal lies, the longer the straight run lasts. In the Carniolan race of the honeybee a straight run lasting a second indi-

cates a target about 500 meters away, and a run lasting two seconds indicates a target two kilometers away. During the dance the follower bees extend their antennae and touch the dancer repeatedly. Within minutes some begin to leave the nest and fly to the target. Their searching is respectably accurate: the great majority come down to search close to the ground within 20 percent of the correct distance.

Superficially the waggle dance of the honeybee may seem to possess some of the more advanced properties of human language. Symbolism occurs in the form of the ritualized straight run, and the communicator can generate new messages at will by means of the symbolism. Furthermore, the target is "spoken of" abstractly: it is an object removed in time and space. Nevertheless, the waggle dance, like all other forms of nonhuman communication studied so far, is severely limited in comparison with the verbal language of human beings. The straight run is after all just a reenactment of the flight the bees will take, complete with wing-buzzing to represent the actual motor activity required. The separate messages are not devised arbitrarily. The rules they follow are genetically fixed and always designate, with a one-to-one correspondence, a certain direction and distance.

In other words, the messages cannot be manipulated to provide new classes of information. Moreover, within this

COURTSHIP RITUAL of grebes is climaxed by the "penguin dance" shown on the following page. In this ritual the male and the female present each other with a beakful of the waterweed that is used as a nest-building material. A pair-bonding display, the dance may have originated as "displacement" behavior, in this instance a pantomime of nest-building triggered by the conflict within each partner between hostility and sexual attraction. The penguin dance was first analyzed in 1914 by Julian Huxley, who observed the ritual among great crested grebes in Europe. Shown here are western grebes in southern Saskatchewan.

rigid context the messages are far from being infinitely divisible. Because of errors both in the dance and in the subsequent searches by the followers, only about three bits of information are transmitted with respect to distance and four bits with respect to direction. This is the equivalent of a human communication system in which distance would be gauged on a scale with eight divisions and direction would be defined in terms of a compass with 16 points. Northeast could be distinguished from north by northeast, or west from west by southwest, but no more refined indication of direction would be possible.

The waggle dance, in particular the duration of the straight run denoting distance, illustrates a simple principle that operates through much of animal communication: the greater the magnitude to be communicated, the more intense and prolonged the signal given. This graduated (or analogue) form of communication is perhaps most strikingly developed in aggressive displays among animals. In the rhesus monkey, for example, a low-intensity aggressive display is a simple stare. The hard look a human receives when he approaches a caged rhesus is not so much a sign of curiosity as it is a cautious display of hostility.

Rhesus monkeys in the wild frequently threaten one another not only with stares but also with additional displays on an ascending scale of intensity. To the human observer these displays are increasingly obvious in their meaning. The new components are added one by one or in combination: the mouth opens, the head bobs up and down, characteristic sounds are uttered and the hands slap the ground. By the time the monkey combines all these components, and perhaps begins to make little forward lunges as well, it is likely to carry through with an actual attack. Its opponent responds either by retreating or by escalating its own displays. These hostile exchanges play a key role in maintaining dominance relationships in the rhesus society.

Birds often indicate hostility by ruffling their feathers or spreading their wings, which creates the temporary illusion that they are larger than they really are. Many fishes achieve the same deception by spreading their fins or extending their gill covers. Lizards raise their crest, lower their dewlaps or flatten the sides of their body to give an impression of greater depth. In short, the more hostile the animal, the more likely it is to attack and the bigger it seems to become. Such exhibitions are often accompanied by graded changes both in color and in vocalization, and even by the release of characteristic odors.

The communication systems of insects, of other invertebrates and of the lower vertebrates (such as fishes and amphibians) are characteristically stereotyped. This means that for each signal there is only one response or very few responses, that each response can be evoked by only a very limited number of signals and that the signaling behavior and the responses are nearly constant throughout entire populations of the same species. An extreme example of this rule is seen in the phenomenon of chemical sex attraction in moths. The female silkworm moth draws males to her by emitting minute quantities of a complex alcohol from glands at the tip of her abdomen. The secretion is called bombykol (from the name of the moth, *Bombyx mori*), and its chemical structure is *trans*-10-*cis*-12-hexadecadienol.

Bombykol is a remarkably powerful biological agent. According to estimates made by Dietrich Schneider and his co-workers at the Max Planck Institute for Comparative Physiology at Seewiesen in Germany, the male silkworm moths start searching for the females when they are immersed in as few as 14,000 molecules of bombykol per cubic centimeter of air. The male catches the molecules on some 10,000 distinctive sensory hairs on each of its two feathery antennae. Each hair is innervated by one or two receptor cells that lead inward to the main antennal nerve and ultimately through connecting nerve cells to centers in the brain. The extraordinary fact that emerged from the study by the Seewiesen group is that only a single molecule of bombykol is required to activate a receptor cell. Furthermore, the cell will respond to virtually no stimulus other than molecules of bombykol. When about 200 cells in each antenna are activated, the male moth starts its motor response. Tightly bound by this extreme signal specificity, the male performs as little more than a sexual guided missile, programmed to home on an increasing gradient of bombykol centered on the tip of the female's abdomen—the principal goal of the male's adult life.

Such highly stereotyped communication systems are particularly important in evolutionary theory because of the possible role the systems play in the origin of new species. Conceivably one small change in the sex-attractant molecule induced by a genetic mutation, together with a corresponding change in the antennal receptor cell, could result in the creation of a population of individuals that would be reproductively isolated from the parental stock. Persuasive

WAGGLE DANCE of the honeybee, first decoded by Karl von Frisch in 1945, is performed by a foraging worker bee on its return to the hive after the discovery of a food source. The pattern of the dance is a repeated figure eight. During the straight run in the middle of the figure the forager waggles its abdomen rapidly and vibrates its wings. As is shown in the illustrations on the opposite page, the direction of the straight run indicates the line of flight to the food source. The duration of the straight run shows workers how far to fly.

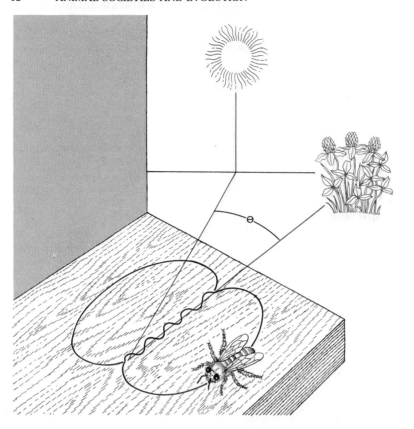

DANCING OUTSIDE THE HIVE on a horizontal surface, the forager makes the straight run of its waggle dance point directly at the source of food. In this illustration the food is located some 20 degrees to the right of the sun. The forager's fellow workers maintain the same orientation with respect to the sun as they leave for the reported source of food.

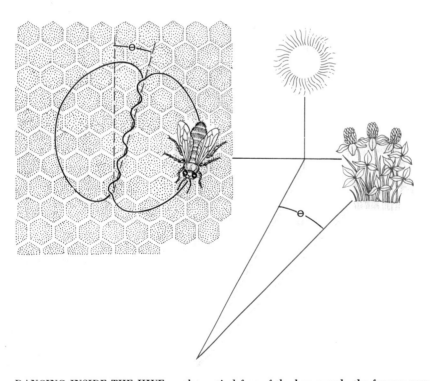

DANCING INSIDE THE HIVE on the vertical face of the honeycomb, the forager uses gravity for orientation. The straight line of the waggle dance that shows the line of flight to the source of food is oriented some 20 degrees away from the vertical. On leaving the hive, the bee's fellow workers relate the indicated orientation angle to the position of the sun.

evidence for the feasibility of such a mutational change has recently been adduced by Wendell L. Roelofs and Andre Comeau of Cornell University. They found two closely related species of moths (members of the genus *Bryotopha* in the family Gelechiidae) whose females' sex attractants differ only by the configuration of a single carbon atom adjacent to a double bond. In other words, the attractants are simply different geometric isomers. Field tests showed not only that a *Bryotopha* male responds solely to the isomer of its own species but also that its response is inhibited if some of the other species' isomer is also present.

A qualitatively different kind of specificity is encountered among birds and mammals. Unlike the insects, many of these higher vertebrates are able to distinguish one another as individuals on the basis of idiosyncrasies in the way they deliver signals. Indigo buntings and certain other songbirds learn to discriminate the territorial calls of their neighbors from those of strangers that occupy territories farther away. When a recording of the song of a neighbor is played near them, they show no unusual reactions, but a recording of a stranger's song elicits an agitated aggressive response.

Families of seabirds depend on a similar capacity for recognition to keep together as a unit in the large, clamorous colonies where they nest. Beat Tschanz of the University of Bern has demonstrated that the young of the common murre (*Uria aalge*), a large auk, learn to react selectively to the call of their parents in the first few days of their life and that the parents also quickly learn to distinguish their own young. There is some evidence that the young murres can even learn certain aspects of the adult calls while they are still in the egg. An equally striking phenomenon is the intercommunication between African shrikes (of the genus *Laniarius*) recently analyzed by W. H. Thorpe of the University of Cambridge. Mated pairs of these birds keep in contact by calling antiphonally back and forth, the first bird vocalizing one or more notes and its mate instantly responding with a variation of the first call. So fast is the exchange, sometimes taking no more than a fraction of a second, that unless an observer stands between the two birds he does not realize that more than one bird is singing. In at least one of the species, the boubou shrike (*Laniarius aethiopicus*), the members of the pair learn to sing duets with each other. They work out combinations of phrases that are sufficiently individual

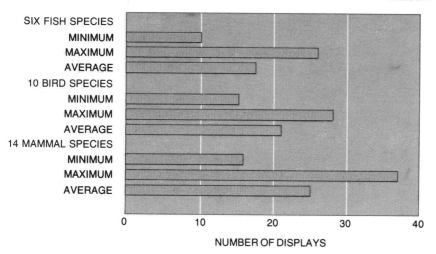

SIX FISH SPECIES
 MINIMUM
 MAXIMUM
 AVERAGE
10 BIRD SPECIES
 MINIMUM
 MAXIMUM
 AVERAGE
14 MAMMAL SPECIES
 MINIMUM
 MAXIMUM
 AVERAGE

0   10   20   30   40

NUMBER OF DISPLAYS

COMMUNICATIVE DISPLAYS used by 30 species of vertebrate animals whose "languages" have been studied vary widely within each of the classes of animals represented: fishes, birds and mammals. The average differences between the classes, however, are comparatively small. The largest and smallest number of displays within each class and the average for each class are shown in this graph. Six of the fish species that have been studied use an average of some 17 displays, compared with an average of 21 displays used by 10 species of birds and an average of 25 displays among 14 species of mammals. Martin H. Moynihan of the Smithsonian Institution compiled the display data. The 30 vertebrates and the number of displays that each uses are illustrated on pages 15 and 16.

to enable them to recognize each other even though both are invisible in the dense vegetation the species normally inhabits.

Mammals are at least equally adept at discriminating among individuals of their own kind. A wide range of cues are employed by different species to distinguish mates, offspring and in the case of social mammals the subordinate or dominant rank of the peers ranged around them. In some species special secretions are employed to impart a personal odor signature to part of the environment or to other members in the social group. As all dog owners know, their pet urinates at regular locations within its territory at a rate that seems to exceed physiological needs. What is less well appreciated is the communicative function this compulsive behavior serves: a scent included in the urine identifies the animal and announces its presence to potential intruders of the same species.

Males of the sugar glider (Petaurus breviceps), a New Guinea marsupial with a striking but superficial resemblance to the flying squirrel, go even further. They mark their mate with a secretion from a gland on the front of their head. Other secretions originating in glands on the male's feet, on its chest and near its arms, together with its saliva, are used to mark its territory. In both instances the odors are distinctive enough for the male to distinguish them from those of other sugar gliders.

As a rule we find that the more highly social the mammal is, the more complex the communication codes are and the more the codes are utilized in establishing and maintaining individual relationships. It is no doubt significant that one of the rare examples of persistent individual recognition among the lower animals is the colony odor of the social insects: ants and termites and certain social bees and wasps. Even here, however, it is the colony as a whole that is recognized. The separate members of the colony respond automatically to certain caste distinctions, but they do not ordinarily learn to discriminate among their nestmates as individuals.

By human standards the number of signals employed by each species of animal is severely limited. One of the most curious facts revealed by recent field studies is that even the most highly social vertebrates rarely have more than 30 or 35 separate displays in their entire repertory. Data compiled by Martin H. Moynihan of the Smithsonian Institution indicate that among most vertebrates the number of displays varies by a factor of only three or four from species to species. The number ranges from a minimum of 10 in certain fishes to a maximum of 37 in the rhesus monkey, one of the primates closest to man in the complexity of their social organization. The full significance of this rule of relative inflexibility is not yet clear. It may be that the maximum number of messages any animal needs in order to be fully adaptive in any ordinary environment, even a social one, is no more than

30 or 40. Or it may be, as Moynihan has suggested, that each number represents the largest amount of signal diversity the particular animal's brain can handle efficiently in quickly changing social interactions.

In the extent of their signal diversity the vertebrates are closely approached by social insects, particularly honeybees and ants. Analyses by Charles G. Butler at the Rothamsted Experimental Station in England, by me at Harvard University and by others have brought the number of individual known signal categories within single species of these insects to between 10 and 20. The honeybee has been the most thoroughly studied of all the social insects. Apart from the waggle dance its known communicative acts are mediated primarily by pheromones: chemical compounds transmitted to other members of the same species as signals. The glandular sources of these and other socially important substances are now largely established. Other honeybee signals include the distinctive colony odor mentioned above, tactile cues involved in food exchange and several dances that are different in form and function from the waggle dance.

Of the known honeybee pheromones the "queen substances" are outstanding in the complexity and pervasiveness of their role in social organization. They include trans-9-keto-2-decenoic acid, which is released from the queen's mandibular glands and evokes at least three separate effects according to the context of its presentation. The pheromone is spread through the colony when workers lick the queen's body and regurgitate the material back and forth to one another. For the substance to be effective in the colony as a whole the queen must dispense enough for each worker bee to receive approximately a tenth of a microgram per day.

The first effect of the ketodecenoic acid is to keep workers from rearing larvae in a way that would result in their becoming new queens, thus preventing the creation of potential rivals to the mother queen. The second effect is that when the worker bees eat the substance, their own ovaries fail to develop; they cannot lay eggs and as a result they too are eliminated as potential rivals. Indirect evidence indicates that ingestion of the substance affects the corpora allata, the endocrine glands that partly control the development of the ovaries, but the exact chain of events remains to be worked out. The third effect of the pheromone is that it acts as a sex attractant. When a virgin queen flies from the hive on her nuptial flight, she releases a vapor trail of the ketodecenoic acid in the air. The smell of the substance not only attracts drones to the queen but also induces them to copulate with her.

Where do such communication codes come from in the first place? By comparing the signaling behavior of closely related species zoologists are often able to piece together the sequence of evolutionary steps that leads to even the most bizarre communication systems. The evolutionary process by which a behavior pattern becomes increasingly effective as a signal is called "ritualization." Commonly, and perhaps invariably, the process begins when some movement, some anatomical feature or some physiological trait that is functional in quite another context acquires a secondary value as a signal. For example, one can begin by recognizing an open mouth as a threat or by interpreting the turning away of the body in the midst of conflict as an intention to flee. During ritualization such movements are altered in some way that makes their communicative function still more effective. In extreme cases the new behavior pattern may be so modified from its ancestral state that its evolutionary history is all but impossible to imagine. Like the epaulets, shako plumes and piping that garnish military dress uniforms, the practical functions that originally existed have long since been obliterated in order to maximize efficiency in display.

The ritualization of vertebrate behavior commonly begins in circumstances of conflict, particularly when an animal is undecided whether or not to complete an act. Hesitation in behavior communicates the animal's state of mind—or, to be more precise, its probable future course of action—to onlooking members of the same species. The advertisement may begin its evolution as a simple intention movement. Birds intending to fly, for example, typically crouch, raise their tail and spread their wings slightly just before taking off. Many species have ritualized these movements into effective signals. In some species white rump feathers produce a conspicuous flash when the tail is raised. In other species the wing tips are flicked repeatedly downward, uncovering conspicuous areas on the primary feathers of the wings. The signals serve to coordinate the movement of flock members, and also may warn of approaching predators.

Signals also evolve from the ambivalence created by the conflict between two or more behavioral tendencies.

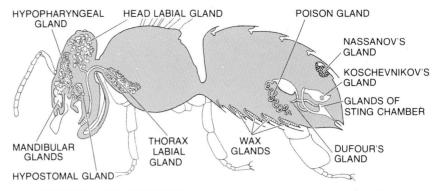

PHEROMONES OF THE HONEYBEE are produced by the glands shown in this cutaway figure of a worker. The glands perform different functions in different castes. In workers, for example, the secretion of the mandibular glands serves as an alarm signal. In a queen, however, the mandibular secretion that is spread through the colony as a result of grooming inhibits workers from raising new queens and also prevents workers from becoming egg-layers. It is also released as a vaporous sex attractant when the new queen leaves the hive on her nuptial flight. The "royal jelly" secreted by the hypopharyngeal gland serves as a food and also acts as a caste determinant. The labial glands of head and thorax secrete a substance utilized for grooming, cleaning and dissolving. Action of the hypostomal-gland secretion is unknown, as is the action of Dufour's gland. The wax glands yield nest-building material, the poison gland is for defense and the sting-chamber glands provide an alarm signal. The secretion of Nassanov's gland assists in assembling workers in conjunction with the waggle dance; that of Koschevnikov's gland renders queens attractive to workers.

When a male faces an opponent, unable to decide whether to attack or to flee, or approaches a potential mate with strong tendencies both to threaten and to court, he may at first make neither choice. Instead he performs a third, seemingly irrelevant act. The aggression is redirected at a meaningless object nearby, such as a pebble, a blade of grass or a bystander that serves as a scapegoat. Or the animal may abruptly commence a "displacement" activity: a behavior pattern with no relevance whatever to the circumstance in which the animal finds itself. The animal may preen, start ineffectual nest-building movements or pantomime feeding or drinking.

Such redirected and displacement activities have often been ritualized into strikingly clear signals. Two classic examples involve the formation of a pair bond between courting grebes. They were among the first such signals to be recognized; Julian Huxley, the originator of the concept of ritualization, analyzed the behavior among European great crested grebes in 1914. The first ritual is "mutual headshaking." It is apparently derived from more elementary movements, aimed at reducing hostility, wherein each bird simply directs its bill away from its partner. The second ritual, called by Huxley the "penguin dance," includes headshaking, diving and the mutual presentation by each partner to its mate of the waterweeds that serve as nesting material. The collection and presentation of the waterweeds may have evolved from displacement nesting

behavior initially produced by the conflict between hostility and sexuality.

A perhaps even more instructive example of how ritualization proceeds is provided by the courtship behavior of the "dance flies." These insects include a large number of carnivorous species of dipterans that entomologists classify together as the family Empididae. Many of the species engage in a kind of courtship that consists in a simple approach by the male; this approach is followed by copulation. Among other species the male first captures an insect of the kind that normally falls prey to empids and presents it to the female before copulation. This act appears to reduce the chances of the male himself becoming a victim of the predatory impulses of the female. In other species the male fastens threads or globules of silk to the freshly captured offering, rendering it more distinctive in appearance, a clear step in the direction of ritualization.

Increasing degrees of ritualization can be observed among still other species of dance flies. In one of these species the male totally encloses the dead prey in a sheet of silk. In another the size of the offered prey is smaller but its silken covering remains as large as before: it is now a partly empty "balloon." The male of another species does not bother to capture any prey object but simply offers the female an empty balloon. The last display is so far removed from the original behavior pattern that its evolutionary origin in this empid species might have remained a permanent mystery if biolo-

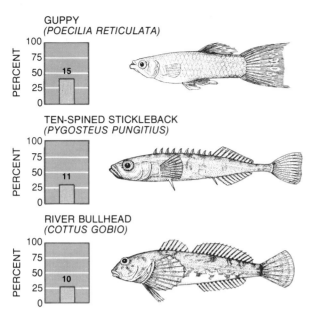
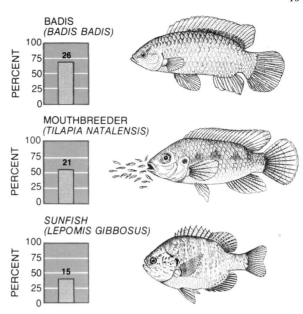

GUPPY
(*POECILIA RETICULATA*)

TEN-SPINED STICKLEBACK
(*PYGOSTEUS PUNGITIUS*)

RIVER BULLHEAD
(*COTTUS GOBIO*)

BADIS
(*BADIS BADIS*)

MOUTHBREEDER
(*TILAPIA NATALENSIS*)

SUNFISH
(*LEPOMIS GIBBOSUS*)

DISPLAYS BY FISHES range from a minimum of 10, used by the river bullhead (*bottom left*), to a maximum of 26, used by the badis (*top right*). The badis repertory is thus more extensive than those of eight of the 10 birds and nine of the 14 mammals studied. The bar beside each fish expressed the number of its displays in percent; 37 displays, the maximum in the study, equal 100 percent.

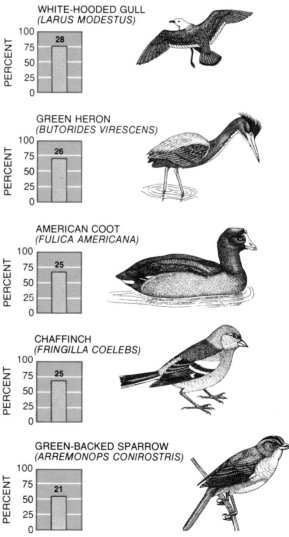

MALLARD DUCK
(*ANAS PLATYRHYNCHOS*)

EASTERN KINGBIRD
(*TYRANNUS TYRANNUS*)

SKUA
(*CATHARACTA SKUA*)

GREAT TIT
(*PARUS MAJOR*)

ENGLISH SPARROW
(*PASSER DOMESTICUS*)

WHITE-HOODED GULL
(*LARUS MODESTUS*)

GREEN HERON
(*BUTORIDES VIRESCENS*)

AMERICAN COOT
(*FULICA AMERICANA*)

CHAFFINCH
(*FRINGILLA COELEBS*)

GREEN-BACKED SPARROW
(*ARREMONOPS CONIROSTRIS*)

DISPLAYS BY BIRDS range from a minimum of 15, used by the English sparrow (*bottom left*), to a maximum of 28, used by the white-headed gull (*top right*). The maximum repertory among birds thus proves to be little greater than the fishes' maximum.

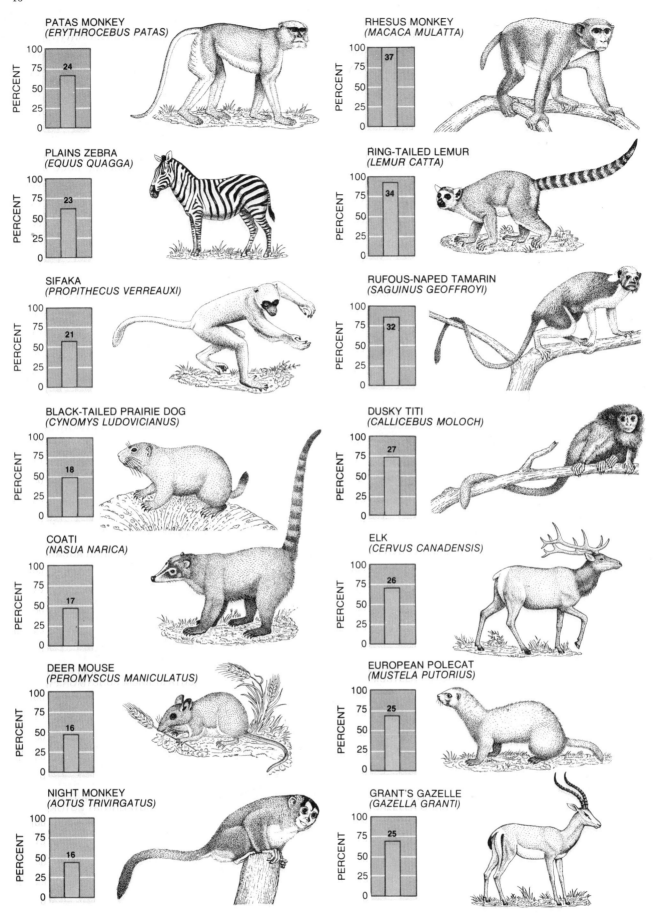

PATAS MONKEY
(ERYTHROCEBUS PATAS)

PERCENT — 24

PLAINS ZEBRA
(EQUUS QUAGGA)

PERCENT — 23

SIFAKA
(PROPITHECUS VERREAUXI)

PERCENT — 21

BLACK-TAILED PRAIRIE DOG
(CYNOMYS LUDOVICIANUS)

PERCENT — 18

COATI
(NASUA NARICA)

PERCENT — 17

DEER MOUSE
(PEROMYSCUS MANICULATUS)

PERCENT — 16

NIGHT MONKEY
(AOTUS TRIVIRGATUS)

PERCENT — 16

RHESUS MONKEY
(MACACA MULATTA)

PERCENT — 37

RING-TAILED LEMUR
(LEMUR CATTA)

PERCENT — 34

RUFOUS-NAPED TAMARIN
(SAGUINUS GEOFFROYI)

PERCENT — 32

DUSKY TITI
(CALLICEBUS MOLOCH)

PERCENT — 27

ELK
(CERVUS CANADENSIS)

PERCENT — 26

EUROPEAN POLECAT
(MUSTELA PUTORIUS)

PERCENT — 25

GRANT'S GAZELLE
(GAZELLA GRANTI)

PERCENT — 25

DISPLAYS BY MAMMALS range from a minimum of 16, used both by the deer mouse and by the night monkey (*left, bottom and next to bottom*), to a maximum of 37, used by the rhesus monkey (*top right*). Two other primates rank next in number of displays.

AGGRESSIVE DISPLAYS by a rhesus monkey (*top*) and a green heron (*bottom*) illustrate a major principle of animal communication: the greater the magnitude to be communicated, the more prolonged and intense the signal is. In the rhesus what begins as a display of low intensity, a hard stare (*left*), is gradually escalated as the monkey rises to a standing position (*middle*) and then, with an open mouth, bobs its head up and down (*right*) and slaps the ground with its hands. If the opponent has not retreated by now, the monkey may actually attack. A similarly graduated aggressive display is characteristic of the green heron. At first (*middle*) the heron raises the feathers that form its crest and twitches the feathers of its tail. If the opponent does not retreat, the heron opens its beak, erects its crest fully, ruffles all its plumage to give the illusion of increased size and violently twitches its tail (*right*). Thus in both animals the likelier the attack, the more intense the aggressive display. Andrew J. Meyerrieck of the University of South Florida conducted the study of heron display and Stuart A. Altmann of the University of Chicago conducted the rhesus display study.

gists had not discovered what appears to be the full story of its development preserved step by step in the behavior of related species.

One of the most important and most difficult questions raised by behavioral biology can be phrased in the evolutionary terms just introduced as follows: Can we hope to trace the origin of human language back through intermediate steps in our fellow higher primates—our closest living relatives, the apes and monkeys—in the same way that entomologists have deduced the origin of the empty-balloon display among the dance flies? The answer would seem to be a very limited and qualified yes. The most probable links to investigate exist within human paralinguistics: the extensive array of facial expressions, body postures, hand signals and vocal tones and emphases that we use to supplement verbal speech. It might be possible to match some of these auxiliary signals with the more basic displays in apes and monkeys. J. A. R. A. M. van Hooff of the State University of Utrecht, for example,

has argued persuasively that laughter originated from the primitive "relaxed open-mouth display" used by the higher primates to indicate their intention to participate in mock aggression or play (as distinct from the hostile open-mouth posture described earlier as a low-intensity threat display in the rhesus monkey). Smiling, on the other hand, van Hooff derives from the primitive "silent bared-teeth display," which denotes submission or at least nonhostility.

What about verbal speech? Chimpanzees taught from infancy by human trainers are reported to be able to master the use of human words. The words are represented in some instances by sign language and in others by metal-backed plastic symbols that are pushed about on a magnetized board. The chimpanzees are also capable of learning rudimentary rules of syntax and even of inventing short questions and statements of their own. Sarah, a chimpanzee trained with plastic symbols by David Premack at the University of California at Santa Barbara, acquired a vocabulary of 128 "words," including a different "name"

for each of eight individuals, both human and chimpanzee, and other signs representative of 12 verbs, six colors, 21 foods and a rich variety of miscellaneous objects, concepts, adjectives and adverbs. Although Sarah's achievement is truly remarkable, an enormous gulf still separates this most intelligent of the anthropoid apes from man. Sarah's words are given to her, and she must use them in a rigid and artificial context. No chimpanzee has demonstrated anything close to the capacity and drive to experiment with language that is possessed by a normal human child.

The difference may be quantitative rather than qualitative, but at the very least our own species must still be ranked as unique in its capacity to concatenate a large vocabulary into sentences that touch on virtually every experience and thought. Future studies of animal communication should continue to prove useful in helping us to understand the steps that led man across such a vast linguistic chasm in what was surely the central event in the evolution of the human mind.

2

The Evolution of Behavior

by John Maynard Smith
September 1978

*Here one of the key questions has to do with altruism:
How is it that natural selection can favor patterns
of behavior that apparently do not favor the survival
of the individual?*

Most species of gulls signal appeasement in fighting by turning their head sharply away from their opponent. This clearly identifiable display is called head flagging. Young gulls do not signal in this way; if they are threatened, they run to cover. One gull species, however, has proved to be an exception to the rule. Chicks of the ledge-nesting kittiwake species do employ the head-flagging display when they are frightened. Their anomalous behavior is the result of the interplay between innate behavior patterns and environmental forces. Unlike other gull species, which live on beaches, the kittiwake perches on tiny ledges of steep cliffs where there is no cover to which the chicks can run if they are threatened. The kittiwake species has responded to environmental pressures by accelerating the development of a standard motor pattern of adult gulls.

This explanation reflects a major change in the understanding of animal behavior. Formerly animal behavior was thought to consist of simple responses, some of them innate and some of them learned, to incoming stimuli. Complex behavior, if it was considered at all, was assumed to be the result of complex stimuli. Over the past 60 years, however, a group of ethologists, notably Konrad Z. Lorenz, Nikolaas Tinbergen and Karl von Frisch, have established a new view of animal behavior. They have shown that the animal brain possesses certain specific competences, that animals have an innate capacity for performing complex acts in response to simple stimuli.

The discovery that certain behavior patterns are inherited was an important contribution to the study of evolution. Genetically determined responses must be subject to the pressures of natural selection. Hence innate behavior must evolve. The ethologists were able to show how a motor pattern employed in a noncommunicatory context such as feeding could evolve into a ritualized form employed as a signal in, say, court-

ship. Differentiation in innate behavior patterns could be traced, as they were in the kittiwake, to selection pressures arising from the environment.

The concept of the evolution of behavior solved some problems and raised others. Since the time of Darwin morphological structures have been used to identify phylogenetic relations. For example, the similarity between a man's arm and a bat's wing is taken as evidence of their common origin. Lorenz pointed out that similarities in behavior patterns can also serve in reconstructing evolutionary history.

It is not always clear, however, how certain types of innate behavior evolved through natural selection. In its modern form the Darwinian interpretation of evolution asserts that (1) evolution consists in changes in the frequency of appearance of different genes in populations and (2) the frequency of the appearance of a particular gene can only increase if the gene increases the "Darwinian fitness" (the expected number of surviving offspring) of its possessors. There are many instances of animal behavior patterns that do not seem to contribute to the survival of the individual displaying that behavior. The classic example is the behavior of the worker bee: this insect will sting an intruder and thereby kill itself in defense of the hive. The problem is evident: How can a gene that makes suicide more likely become established?

The concern over this type and other types of apparently anomalous behavior led to the development of a new phase in the study of the evolution of behavior: a marriage of ethology and population genetics. From this perspective it has been possible to explain how natural selection operates to bring about the evolution of many of the most perplexing examples of animal behavior. In this article I shall discuss the progress that has been made in understanding the evolution of two important types of behavior: cooperative or altruistic behavior such as that of the worker bee

and ritualized behavior in animal contests. I shall begin with the first type of behavior, one of the initial problems to which the new discipline was successfully applied.

For a long time many biologists, particularly those unfamiliar with genetics, explained the evolution of behavior such as the altruism of the worker bee by saying that this type of behavior contributed to the "good of the species." A behavior pattern that promoted the survival of a species would, they believed, be favored by natural selection even if it reduced the Darwinian fitness of the individual displaying it. There is an obvious problem with that explanation: if a gene increases the fitness of an individual, then it will be established in a species even if it reduces the long-term survival of the species.

Darwin and later the founders of population genetics, R. A. Fisher, Sewall Wright and J. B. S. Haldane, were aware of the problem and even came close to solving it. The current understanding of the evolution of altruistic behavior, however, is due to the work of W. D. Hamilton of the Imperial College of Science and Technology in London. Hamilton presented his thesis in two papers published in 1964. To understand Hamilton's argument, consider the fact that a parent may risk its life in defense of its offspring, say by feigning injury to distract a predator. In this way the parent may increase its own Darwinian fitness. Although it is possible that both the parent and the offspring will be killed, it is more likely that both the parent and the offspring will survive. In the latter case the parent's Darwinian fitness will be greater after the altruistic act than it would have been if the parent had left its offspring to the predator. The genes associated with the altruistic act (in this instance feigning injury) may be present in the offspring, so that their frequency may be increased. Hence natural selection favors parental altruism, that is, it is through parental altruism that the par-

DIFFERENTIATION in innate behavior patterns is the result of selection pressures arising from the environment. For example, adult gulls signal appeasement in fighting with a standard head-flagging display, turning their head sharply away from an opponent (*top*). Most young gulls do not employ the display; if they are threatened, they run to cover. Chicks of the ledge-nesting kittiwake species, however, do employ the head-flagging display when they are threatened (*bottom*). Unlike other gull species, which live on beaches, the kittiwake lives on tiny ledges of steep cliffs where there is no cover to which the chicks can run. Early development of the head-flagging behavior pattern contributes to the survival of the gulls. Hence natural selection favors the evolution of the kittiwake's anomalous behavior.

ent's behavioral characteristic is established in future generations. Hamilton realized that this analysis of parental altruism could also be applied to explain acts increasing the chances of survival of relatives other than children, for example siblings or even cousins. It was this basic perception that was the key to understanding the evolution of a wide range of animal behavior.

Consider two individuals, a "donor" and a "recipient." The donor performs an act that reduces its own Darwinian fitness, or expected number of surviving offspring, by a cost C but increases the recipient's fitness by a benefit B. Suppose that there is a pair of allelic, or alternative, genes A and a and that the

presence of A makes an individual more likely to perform the act. Hamilton showed that the change in the frequency of gene A in the population after the act depends on the coefficient of relationship r between the donor and the recipient, that is, on the average fraction of genes of common descent in individuals with the genetic relationship of the donor and the recipient. More precisely, he showed that (with certain approximations) the frequency of gene A will increase because of the donor's act if the coefficient of relationship r is greater than C/B.

For example, if an individual has two sets of genes, one from a father with two sets and one from a mother with

two sets, then on the average the probability is 1/2 that any particular gene in the individual is present in a full sibling [see illustration on this page]. Hence the coefficient of relationship between the individual and its full sibling is 1/2. Therefore, according to Hamilton's argument, if a gene in the individual causes it to sacrifice its life to save the lives of more than two siblings, then the number of replicas of the gene present after the sacrifice is made is greater than the number present would be if the sacrifice had not been made. The sacrifice is selectively advantageous. (In this instance the cost C is equal to 1 and the benefit B is equal to more than 2, so that the coefficient of relationship 1/2 is indeed greater than C/B.)

Hamilton's work predicts that altruistic and cooperative behavior will be found more frequently in interactions of related individuals than in interactions of unrelated individuals. Observation certainly confirms this prediction. In fact, as Hamilton points out, the highest degree of cooperation is displayed by colonies of genetically identical cells such as the cells that make up the human body. It is important to note that these concepts apply to organisms incapable of recognizing degrees of relationship. In species that usually live in family groups a gene causing an individual to act altruistically toward members of its community will increase in frequency even if the individuals carrying it cannot recognize family members. The warning signals given by birds and mammals (such as a rabbit's thumping the ground with its hind legs) exemplify this type of altruistic behavior.

One of the best illustrations of the same type of altruism is the behavior of a viruslike self-replicating particle called a plasmid, which lives as a parasite in bacteria. On occasion a plasmid manufactures a toxin that kills the host bacterium and probably the plasmid as well. When the host dies, the toxin is released, but it kills only those nearby bacteria that do not harbor plasmids. The bacteria with plasmids are unharmed because each plasmid also manufactures an immunity protein that protects it against the toxin of other plasmids. Therefore by killing off competing bacteria the suicidal toxin-producing gene contributes to the survival of those bacteria that carry its genetic replicas. This interpretation is supported by the fact that the plasmids tend to produce toxin when bacteria are crowded and in competition.

The most unexpected demonstration of the strength of Hamilton's theory was his use of it to explain the evolution of the social insects. Such insects live in advanced social orders characterized by cooperation, caste specialization and in-

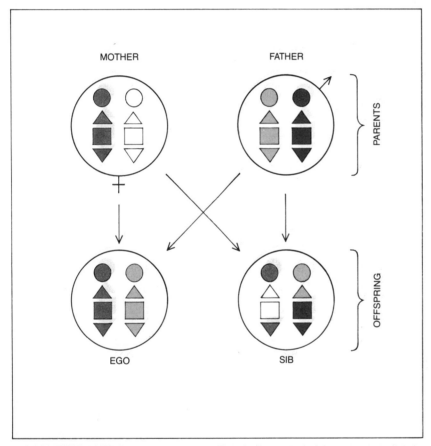

ALTRUISTIC ACTS do not appear to contribute to the survival of the animals performing them, but their evolution can be understood by examining the genetic relationship between the performer of an act and the beneficiary of the act. The genetic relationship between any two individuals can be expressed by a coefficient of relationship that is defined as the average fraction in the individuals of shared genes, or genes of common descent. For example, this illustration shows two parents, each with two sets of four genes, and two offspring: "Ego" and a full sibling, "Sib." The four pairs of alleles (alternative forms of the same gene) in each individual are represented by four different shapes. The color of the genes in the offspring indicates the manner in which parental alleles have been reassorted. Ego and Sib each have two sets of inherited genes, one from their mother and one from their father, so that on the average the probability is 1/2 that a gene present in Ego is also present in Sib. Hence the coefficient of relationship between the two siblings is 1/2. Modern evolutionary theory states that evolution consists in changes in the frequency of the appearance of various genes in a population and that a gene can only increase in frequency if it increases the Darwinian fitness, or expected number of surviving offspring, of its possessor. W. D. Hamilton of the Imperial College of Science and Technology in London showed that (with certain approximations) the gene associated with an altruistic act will increase in frequency because of the act only if the coefficient of relationship between the performer and the beneficiary is greater than C/B, where C is the cost (in Darwinian fitness) of the act to performer and B is benefit (in Darwinian fitness) of act to beneficiary.

dividual altruism. Fully social insects (insects that have all three social characteristics) exhibit a reproductive division of labor, with more or less sterile individuals (workers) laboring on behalf of fecund individuals (queens). With the single exception of the termites, all fully social insect species belong to the order Hymenoptera. The order also includes many nonsocial species, and the surprising fact is that sociality has originated on a number of separate occasions among the bees, the ants and the wasps. Hamilton was able to trace the predisposition for sociality to a particular feature of the genetic system of this order of insects.

In the hymenoptera females develop from fertilized eggs and so are diploid: they have two sets of chromosomes. Males develop from unfertilized eggs and so are haploid: they have a single set of chromosomes. In a population where both sexes are diploid, such as the one described above, the coefficient of relationship, or the average fraction of shared genes, between a mother and a daughter is the same as the coefficient of relationship between any two full siblings: in both instances r equals $1/2$. As a result of the haplo-diploidy of the hymenoptera, however, a female has more genes in common with her full sisters than she has with her own daughters. Each female receives half of her genes from her haploid father and half from her diploid mother. Sisters share all of the genes they receive from their father (because he has only one set) and on the average half of the genes they receive from their mother (because she has two sets). Hence the coefficient of relationship between a mother and a daughter is still $1/2$ but the coefficient of relationship between full sisters is $(1/2) \times (1) + (1/2) \times (1/2)$, or $3/4$ [*see illustration on next page*].

To grasp the import of these numbers consider a species in which a female constructs and provisions a nest cell for each of her eggs and continues to lay eggs past the time when her first daughter reaches maturity. All fully social insects display an overlap of this kind. The coefficients of relationship indicate that, other things being equal, the daughter will do the most to perpetuate her genes if instead of leaving and provisioning nest cells containing her own daughters, she stays with her mother and provisions cells containing her sisters. In this way the genetic makeup of the hymenoptera predisposes them to evolve a social system in which sterile female workers care for their full siblings.

There is another twist to this argument that is still being discussed. For a female hymenopteran the coefficient of relationship is $1/2$ with a son and $1/4$ with a brother. Hence although

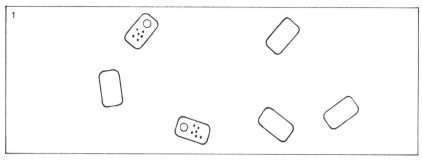
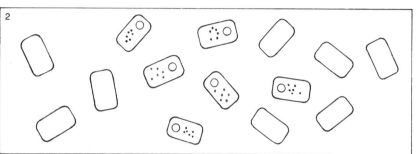
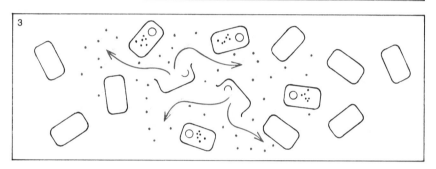
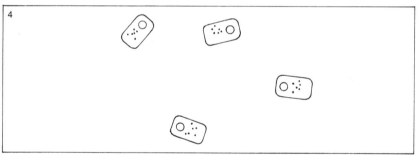
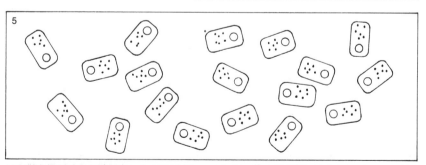

ALTRUISM OF A KIND is displayed by a viruslike self-replicating particle known as a plasmid (*open circles*), which lives as a parasite in a bacterium (*1*). Each plasmid manufactures an immunity protein (*black dots*). When the bacteria become overcrowded, some of the plasmids manufacture a colicin, or toxin (*color dots*), that kills their host bacteria and probably the plasmids themselves (*2*). When host bacteria die, colicin is released (*3*), thereby killing all nearby bacteria that do not contain the immunity protein and leaving only bacteria that are host to plasmids (*4*). The plasmids that produced colicin are destroyed, but their genetic replicas are able to multiply without competition (*5*). As example demonstrates, selection on the basis of genetic relationship operates even when individuals cannot recognize degrees of relationship.

she should raise sisters in preference to daughters, theoretically she should raise sons in preference to brothers. Hamilton points out that in fact in many social species the workers lay unfertilized eggs and raise these sons in preference to their brothers.

The matter has been taken a step further by Robert L. Trivers and H. Hare of Harvard University. If a female hymenopteran cannot distinguish between male and female eggs, she is obliged to allocate her time equally to males and females. She will therefore gain as much by raising her own offspring (r equals $1/2$ for sons and r equals $1/2$ for daughters) as by raising her full siblings (r equals $1/4$ for brothers and r equals $3/4$ for sisters). Trivers and Hare point out that in cases where the workers can distinguish the sex of the larvae they are raising they should raise an excess of females (r equals $3/4$ for sisters) over males (r equals $1/4$ for brothers). In

fact, they showed that if the sex ratio among the reproducing members of the colony is determined by the genes in the workers, it will be about three females for every male, whereas if it is determined by genes in the queen, it will be about one female for every male. Analysis of data on ants indeed shows an excess investment in the biomass of females at the rate of about three to one. Therefore it can be deduced that the sex ratio in ants is controlled by the workers rather than by the queen. With this distorted sex ratio the worker ants should care for their siblings, male and female, in preference to their own offspring.

In recent years Hamilton's ideas have been increasingly applied to the study of the social life of higher animals, in particular birds and mammals. One such application involves the many primate species that live in groups consisting of several adult males, several adult

females and their young. It is becoming apparent that in these species the young of one sex, usually the males, leave their natal troop when they reach sexual maturity and join another troop to breed. Craig Packer of the University of Sussex encountered this behavior pattern in his study of three troops of olive baboons (*Papio anubis*) in the Gombe National Reserve in Tanzania. Of 41 intertroop transfers observed over a six-year period 39 involved males, and all the males that reached maturity during that time left their natal troop.

It appears that young males leave their troop because the females in it refuse to mate with them and because the young males are attracted to unfamiliar females. This behavior seems to be selectively advantageous because a male and female born in the same troop are often close relatives and so would produce inbred offspring of low fitness. For the purposes of the present discussion,

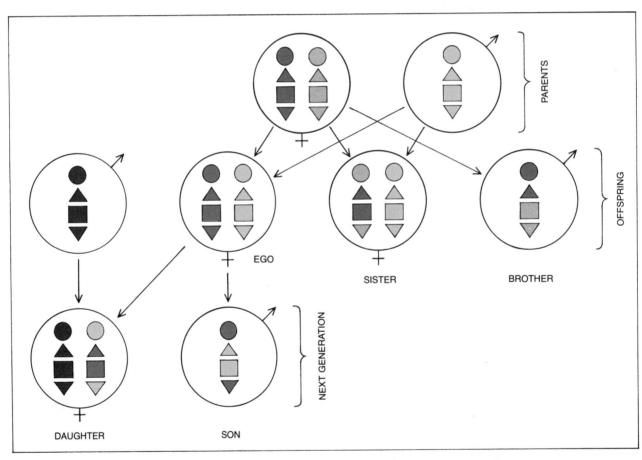

FULLY SOCIAL INSECT SPECIES are characterized by cooperation, caste specialization and individual altruism. With the exception of the termites, they all belong to the order Hymenoptera. Hamilton analyzed the frequent evolution of social behavior among these insects by examining the genetic structure of the order: females develop from fertilized eggs and have two sets of chromosomes but males develop from unfertilized eggs and have only one set of chromosomes. Consider the female Ego shown in this illustration. Any female inherits her two sets of (possibly reassorted) genes: one from her mother, with two sets, and one from her father, with one set. Hence the coefficient of relationship (average fraction of shared genes) between Ego and a full sister Sib is $(1/2) \times (1/2) + (1/2) \times (1)$, or $3/4$, but the coefficient of relationship between Ego and a daughter is $1/2$. Ego has more genes in common with her sister than with her daughter. If Ego's mother continues to provision cells for eggs after Ego reaches maturity (all social insects display a reproductive overlap of this type), then Ego will do the most to perpetuate her genes if she helps to provision cells containing her sisters rather than provisioning cells containing her daughters. Genetic makeup of hymenoptera predisposes them to evolve social system in which sterile female workers care for siblings.

however, the most interesting aspect of this behavior pattern is not its genetic causes but its genetic effects, that is, its consequences for the genetic relationships within the troop.

As a result of the transfers the females in an olive-baboon troop will be closely related but the adult breeding males will in general not be related. In chimpanzee troops, where the males form the permanent basis of the troop and the females transfer, the situation is reversed. According to Hamilton's thesis, strong cooperation can be expected among female baboons and among male chimpanzees but not among male baboons or among female chimpanzees. The validity of this prediction is still a matter of controversy. My own guess is that it will eventually be upheld.

In spite of male baboon's lack of genetic relationship, they do display one type of cooperative behavior. When two baboons are in some kind of contest, one of them may enlist the aid of a third baboon. The soliciting baboon asks for help with an easily recognized signal, turning its head repeatedly back and forth between its opponent and its potential assistant. Packer recorded 140 instances of this type of behavior. Twenty of them involved a male A soliciting the help of another male B to take over an estrous female that was consorting with a third male C. In six of these instances the attempt was successful, and each time it was the soliciting male A that obtained the female. The fitness of the assisting male B does not appear to be increased by this behavior, and so the obvious question is: What does B get in return for its services?

The most convincing explanation of the evolution of this type of altruistic behavior between unrelated individuals is found in the concept of reciprocal altruism formulated by Trivers. According to Trivers' hypothesis, male B, by helping male A (without incurring great personal risk), gains the assurance that on a future occasion A will help it in return. Hence male B is most probably increasing its own Darwinian fitness, and the gene giving rise to this type of altruistic behavior will probably increase in frequency.

One problem with this explanation is that there appears to be no defense against cheating. What prevents male A from accepting help but later refusing to reciprocate? The answer may be that baboon behavior patterns have evolved so that the animals help only those individuals that do reciprocate. In that case cheating would not pay. Of course, this hypothesis presupposes individuals can recognize other individuals and remember their past behavior, but it is quite reasonable to assume that baboons possess such capabilities. Packer's data certainly support this assumption. He

Suppose that a population contains a small fraction p of behavioral "mutants" adopting strategy J and that the remainder of the population q adopts strategy I. If the total Darwinian fitness (expected number of surviving offspring) of the members of the population before a series of contests is C, then after the contests

$$W(I) = C + qE(I,I) + pE(I,J) \quad \text{and}$$
$$W(J) = C + qE(J,I) + pE(J,J),$$

where $E(I,J)$ is the expected payoff (change in fitness) to an individual employing strategy I in a contest with an individual employing strategy J, $W(I)$ is the total increased fitness acquired by employing strategy I and so on.

If I is an evolutionarily stable strategy, then $W(I) > W(J)$ for any mutant strategy J. In this case

$$\text{either } E(I,I) > E(J,I) \quad \text{or}$$
$$E(I,I) = E(J,I) \quad \text{and} \quad E(I,J) > E(J,J).$$

GAME-THEORY MODELS help to explain the use of conventions in animal contests, another type of behavior that does not seem to promote the survival of the individual displaying it. For each model there is sought an evolutionarily stable strategy, that is, a strategy that confers the highest reproductive fitness on the animals adopting it. This illustration shows the mathematical requirements for a strategy to be evolutionarily stable. More generally, an evolutionarily stable strategy can be defined as a strategy with the property that if all members of a population adopt it, then no mutant strategy can invade the population. It is assumed that members of model population engage in contests in random pairs and that subsequently each individual reproduces in proportion to payoff (change in Darwinian fitness) it has accumulated. It now appears that many types of conventional fighting are indeed evolutionarily stable strategies.

found that the male baboons responding most frequently to solicitations for aid also received aid most frequently and that males tended to solicit the aid of particular partners that in turn solicited aid from them.

Over the past few years I have become particularly interested in the evolution of a type of ritualized animal behavior: the use of conventions in animal contests. Animals engaged in a contest over some valuable resource (such as a mate, territory or position in a hierarchy) do not always use the weapons available to them in the most effective way. They may instead act according to certain conventions (employing threat displays, refraining from attacking an opponent in a vulnerable position and so on), often pursuing a kind of limited warfare that avoids serious injury. For example, when male fiddler crabs fight over the possession of a burrow, they use a powerful enlarged claw as a weapon. Although the claw is strong enough to crush the abdomen of an opponent, no crab has ever been known to injure another in such a fight. (It would be wrong to deduce from this example that animals are never injured in intraspecific fights or that animals never fight to the death. The conventional behavior is sufficiently common, however, to require an explanation.)

At the time when I first learned of the problem the evolution of conventional fighting was explained by arguing that if intraspecific fighting were not conventional, then a great many animals would be injured. In other words, conventional behavior evolved because unconventional behavior would, as Julian Huxley had put it, "militate against the survival of the species." As a student of Hal-

dane's I had been taught to be distrustful of arguments that depend on "the good of the species." This particular one did not seem to be able to account for the complex anatomical and behavioral adaptations for limited conflict found in many species. I thought there should be a way to explain how natural selection operates on the individual to promote those characteristics, that is, to show that conventional behavior increases the Darwinian fitness of the individual displaying it.

It appeared, however, that an individual's fitness would be increased not by conventional fighting but by unconventional fighting. It seemed to me that, in a contest between individuals A and B, if A obeyed the rules and B "hit below the belt," then B would win the contest and pass its genes on to the next generation. This puzzle remained at the back of my mind until 1970, when an unpublished paper by G. R. Price prompted me to review it. It occurred to me then that I might gain some understanding of the problem by borrowing some of the concepts of the mathematical theory of games.

Game theory was formulated by John von Neumann and Oskar Morgenstern in the 1940's for the purpose of analyzing human conflict. The theory in particular seeks to determine the optimum strategy to pursue in conflict situations. I hoped that by applying a modified form of game theory I would be able to construct a mathematical model of animal contests and so determine what strategies would be favored by natural selection at the level of the individual. If all went well, experimental evidence and observation would support the mathematically derived conclusions.

The strategies I was seeking had little

to do with the optimum strategies with which traditional game theory deals. For each game model of animal contests I hoped to determine an evolutionarily stable strategy: a strategy with the property that if most of the members of a large population adopt it, then no mutant strategy can invade the population. In other words, a strategy is evolutionarily stable if there is no mutant strategy that gives higher Darwinian fitness to the individuals adopting it.

Consider a simple model: a species that in contests between two individuals has only two possible tactics, a "hawk" tactic and a "dove" one. A hawk fights without regard to any convention and escalates the fighting until it either wins (that is, until its opponent runs away or is seriously injured) or is seriously injured. A dove never escalates; it fights conventionally, and then if its opponent escalates, it runs away before it is injured.

At the end of a contest each contestant receives a payoff. The expected payoff to individual X in a contest with individual Y is written $E(X,Y)$. The payoff is a measure of the change in the fitness of X as a result of the contest, and so it is determined by three factors: the advantage of winning, the disadvantage of being seriously injured and the disadvantage of wasting time and energy in a long contest. For the hawk-dove game suppose the effect on individual fitness is $+10$ for winning a contest and -20 for suffering serious injury. Suppose further

two doves can eventually settle a contest but only after a long time and at a cost of -3. (The exact values of the payoffs do not affect the results of the model as long as the absolute, or unsigned numerical, value of injury is greater than that of victory.)

The game can be analyzed as follows. If the two individuals in a contest both adopt dove tactics, then since doves do not escalate, there is no possibility of injury and the contest will be a long one. Each contestant has an equal chance of winning, and so the expected payoff to one of the doves D equals the probability of D winning the contest (p equals $1/2$) times the value of victory plus the cost of a long battle, that is, $E(D,D)$ equals $(1/2)(+10) + (-3)$, or $+2$. Similarly, a hawk fighting another hawk has equal chances of winning or of being injured but in any case the contest will be settled fairly quickly. Hence the expected payoff $E(H,H)$ is equal to $(1/2)(+10) + (1/2)(-20)$, or -5. A dove fighting a hawk will flee when the hawk escalates, so that the dove's expected payoff is 0 and the victorious hawk's payoff is $+10$.

Now suppose the members of a population engage in contests in the hawk-dove game in random pairs and subsequently each individual reproduces its kind (individuals employing the same strategy) in proportion to the payoff it has accumulated. If there is an evolutionarily stable strategy for the game, the population will evolve toward it. The question, then, is: Is there an evolu-

tionarily stable strategy for the hawk-dove game?

It is evident that consistently playing hawk is not an evolutionarily stable strategy: a population of hawks would not be safe against all mutant strategies. Remember that in a hawk population the expected payoff per contest to a hawk $E(H,H)$ is -5 but the expected payoff to a dove mutant $E(D,H)$ is 0. Hence dove mutants would reproduce more often than hawks. A similar argument shows that consistently playing dove is also not an evolutionarily stable strategy.

There is a precise mathematical definition for an evolutionarily stable strategy: A strategy I is evolutionarily stable if, for any mutant strategy J, either $E(I,I)$ is greater than $E(J,I)$ or $E(I,I)$ equals $E(J,I)$ and $E(I,J)$ is greater than $E(J,J)$. Although neither of the pure strategies labeled "Always play hawk" or "Always play dove" fulfills one of these requirements, there is a mixed strategy that does. A mixed strategy is one that prescribes different tactics to be followed in a game according to a specified probability distribution. The mixed strategy that is evolutionarily stable for the hawk-dove game is play hawk with probability $8/13$ and play dove with probability $5/13$. I shall not discuss the derivation of this strategy here, but it is not difficult to see that the strategy does fulfill the second requirement of being evolutionarily stable against, say, a mutant hawk strategy.

If the mixed strategy is called M, it will suffice to show that $E(M,M)$ equals $E(H,M)$ and $E(M,H)$ is greater than $E(H,H)$. This can be done by applying the definition of strategy M: the payoff $E(M,M)$ is equal to $(8/13)E(H,M) + (5/13)E(D,M)$, and $E(H,M)$ is equal to $(8/13)E(H,H) + (5/13)E(H,D)$ and $E(D,M)$ is equal to $(8/13)E(D,H) + (5/13)E(D,D)$. The values already computed for the hawk-dove game can now be substituted into these equations so that $E(M,M)$ is equal to $(8/13)(10/13) + (5/13)(10/13)$, or $10/13$. The preceding calculation showed that $E(H,M)$ equals $10/13$, and so $E(M,M)$ and $E(H,M)$ are equal. Furthermore, the payoff $E(M,H)$ equals $(8/13)E(H,H) + (5/13)E(D,H)$, or $-40/13$, and $E(H,H)$ equals -5, and so $E(M,H)$ is greater than $E(H,H)$. In other words, the strategy hawk cannot invade a population employing the mixed strategy M.

The hawk-dove model predicts that mixed strategies will be found in real animal contests, either in the form of different animals adopting different tactics (such as hawk and dove) or in the form of individuals varying their tactics. Animal behavior in many contest situations is indeed variable, but of course that does not prove that an evolutionari-

$E(H,H) = \frac{1}{2}(+10) + \frac{1}{2}(-20) = -5$

$E(H,D) = +10$

$E(D,H) = 0$

$E(D,D) = \frac{1}{2}(+10) + (-3) = +2$

SERIOUS INJURY $= -20$

VICTORY $= +10$

LONG CONTEST $= -3$

	HAWK (H)	DOVE (D)
HAWK (H)	-5	$+10$
DOVE (D)	0	$+2$

IN THE HAWK-DOVE GAME, illustrated here, there are only two tactics that can be employed in contests between two individuals: a "hawk" tactic and a "dove" one. A hawk fights without regard to any convention and escalates a contest until it wins or is seriously injured. A dove fights conventionally, never escalating; if its opponent escalates, it runs away before it is injured; two doves can settle a contest but only after a long period of time. The changes in a contestant's Darwinian fitness as a result of serious injury, of a long contest and of victory are shown at the upper left in the illustration. (The exact values of these factors do not affect the model results as long as the unsigned numerical value of injury is greater than that of victory.) Calculations of the expected payoffs to individuals in different contests are shown at the upper right. The payoffs are displayed in the matrix at the bottom. Each payoff is to the individual employing the tactic directly to the left in the matrix in a contest with an individual employing the tactic directly above. For example, $+10$ (color) equals $E(H,D)$, the expected payoff to a hawk H in contest with a dove D. In hawk-dove game neither of pure strategies designated "Always play hawk" or "Always play dove" is evolutionarily stable. Only evolutionarily stable strategy is mixed strategy: play hawk with probability 8/13 and play dove with probability 5/13.

ly mixed strategy is operating. One case of animal behavior that does conform rather well to the model is found in investigations into the behavior of the dung fly conducted by G. A. Parker of the University of Liverpool.

Female dung flies lay their eggs in cowpats, and so males congregate at cowpats and try to mate with the arriving females. Parker found that the rate at which the females arrive at a cowpat decreases as the cowpat gets stale. In game terms the male is presented with a choice of two tactics as the cowpat he is patrolling gets stale. He can leave in search of a fresh cowpat or he can stay. The success of the male's choice of tactic of course depends on the behavior of other males. If most of the other males leave as soon as the cowpat gets stale, then he should stay, because although relatively few females will be arriving, he will have little or no competition in mating with them. On the other hand, if the other males stay, then he should leave. In other words, the only evolutionarily stable strategy is a mixed one in which some males leave early and others stay. Game-theory analysis predicts that with this strategy when the system reaches an equilibrium, early-leaving and late-leaving males should have the same average mating success. Parker's data yield precisely that result. It is not known, however, whether the evolutionarily stable mixed strategy of the dung fly is achieved by some males' consistently leaving early and others' consistently leaving late or by individual males' varying their tactics.

It is obvious that real animals can adopt strategies that are more complex than "Always escalate," "Always display" or some mixture of the two. For example, some animals make probes, or trial escalations. Others employ conventional tactics but will escalate in retaliation for an opponent's escalation. There is, however, another important way in which many real animal contests do not conform to the hawk-dove model. Most real contests are asymmetric in that, unlike hawks and doves, the contestants differ from each other in some area besides strategy.

Three basic types of asymmetries are encountered in animal contests. First, there are asymmetries in the fighting ability (the size, strength or weapons) of the contestants; differences of this kind are likely to affect the outcome of an escalated fight. Second, there are asymmetries in the value to the contestants of the resource being competed for (as in a contest over food between a hungry individual and a well-fed one); differences of this kind are likely to affect the payoffs of a contest. Third, there are asymmetries that are called uncorrelated because they affect neither the outcome of

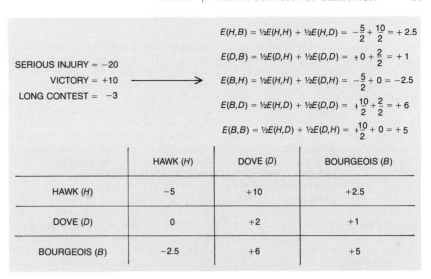

$$E(H,B) = \tfrac{1}{2}E(H,H) + \tfrac{1}{2}E(H,D) = -\tfrac{5}{2} + \tfrac{10}{2} = +2.5$$

$$E(D,B) = \tfrac{1}{2}E(D,H) + \tfrac{1}{2}E(D,D) = +0 + \tfrac{2}{2} = +1$$

$$E(B,H) = \tfrac{1}{2}E(H,H) + \tfrac{1}{2}E(D,H) = -\tfrac{5}{2} + 0 = -2.5$$

$$E(B,D) = \tfrac{1}{2}E(H,D) + \tfrac{1}{2}E(D,D) = +\tfrac{10}{2} + \tfrac{2}{2} = +6$$

$$E(B,B) = \tfrac{1}{2}E(H,D) + \tfrac{1}{2}E(D,H) = +\tfrac{10}{2} + 0 = +5$$

SERIOUS INJURY = −20
VICTORY = +10
LONG CONTEST = −3

	HAWK (H)	DOVE (D)	BOURGEOIS (B)
HAWK (H)	−5	+10	+2.5
DOVE (D)	0	+2	+1
BOURGEOIS (B)	−2.5	+6	+5

HAWK-DOVE-BOURGEOIS GAME, illustrated here, models animal contests that are characterized by uncorrelated asymmetries, that is, differences between contestants that do not necessarily affect the outcome or payoffs of the contests. Asymmetries of this type often serve to settle real contests conventionally. A contest over some resource between the owner of the resource and an interloper is a good example of an uncorrelated asymmetry, and so it was used to define a new tactic, "bourgeois," to be added to the tactics in the hawk-dove game. If a bourgeois contestant is the owner of the resource in question, it adopts the hawk tactic; otherwise it adopts the dove tactic. It is assumed that each contest is between an owner and an interloper, that each individual is equally likely to be in either role and that each individual knows which role it is playing. Pure strategy "bourgeois" is the only evolutionarily stable strategy for the game. There can never be an escalated contest between opponents adopting strategy, since one will be owner and playing hawk and other will be interloper and playing dove. Hence ownership is taken as conventional cue for settling contests in model population. Many examples of bourgeois strategy have been found in real animal populations (*see illustration on next two pages*).

escalation nor the payoffs of a contest. For the purposes of this discussion the uncorrelated asymmetries are of special interest because they often serve to settle contests conventionally.

Perhaps the best example of an uncorrelated asymmetry is found in a contest over a resource between the "owner" of the resource and an interloper. In calling this an uncorrelated asymmetry I do not mean that ownership never alters the outcome of escalation or the payoffs of contests; I simply mean that ownership will serve to settle contests even when it does not alter those factors. To demonstrate the effect of such an uncorrelated asymmetry I shall return to the hawk-dove game and add to it a third strategy called bourgeois: if the individual is the owner of the resource in question, it adopts the hawk tactic; otherwise it adopts the dove tactic.

In the hawk-dove-bourgeois game it is assumed that each contest is between an owner and an interloper, that each individual is equally likely to be in either role and that each individual knows which role it is playing. The payoffs for contests involving hawks and doves are unchanged by the addition of the new strategy, but additional payoffs must be calculated for contests that involve bourgeois contestants [*see illustration above*]. For example, in a contest between a bourgeois and a hawk there

is an equal chance that the bourgeois will be the owner (and so playing hawk) or the interloper (and so playing dove); hence $E(B,H)$ equals $(1/2)E(H,H) + (1/2)E(D,H)$, or $(1/2)(-5) + (1/2)(0)$, or -2.5. The remaining payoffs are calculated in a similar manner. The main point, however, is that there can never be an escalated contest between two opponents playing bourgeois, because if one is the owner and playing hawk, then the other is the interloper and playing dove. Therefore the payoff $E(B,B)$ is equal to $(1/2)E(H,D) + (1/2)E(D,H)$, or $(1/2)(10) + (1/2)(0)$, or 5. When this figure is compared with the other payoffs, it is not difficult to see that consistently playing bourgeois is the only evolutionarily stable strategy for this game. Thus ownership is taken as a conventional cue for settling contests.

Hans Kummer of the University of Zurich has observed a beautiful example of the bourgeois strategy in the hamadryas baboon (*Papio hamadryas*). In this species a single male forms a permanent bond with one or more females and is not normally challenged by other males. Kummer performed the following experiment with three unacquainted baboons. Male *A* and a female were put in an enclosure and male *B* was put in a cage from which he could see what was happening in the enclosure but could not interfere. In a relatively short time (about 20 minutes) a bond was established between male *A* and the female. Male *B* was then released into the enclosure. He did not attempt to annex the female and in fact avoided any kind of confrontation with *A*.

There are two possible explanations for male *B*'s behavior. It may be that, as the model predicts, ownership is taken as a conventional cue for the settlement of contests. On the other hand, male *B* may have perceived that male *A* was stronger and would probably have won an escalated contest. Kummer was able to eliminate the second possibility by repeating the experiment with the same two males and a different female several weeks later. Now, however, the roles of the male baboons were reversed and *B* was placed in the enclosure with the female and *A* was placed in the cage. This time it was *B* that annexed the female

and was not challenged by *A*. The bourgeois principle was indeed operating. (It should be noted that there is more to the story; since these experiments were done Kummer has found that female preference also plays a role.)

N. B. Davies of the University of Oxford has discovered another example of the bourgeois strategy in the speckled wood butterfly (*Pararge aegeria*). Males of the species claim and defend sunlit spots on the forest floor, where they can court more females than they can in the forest canopy. There are never enough spots for all the males to occupy at any one time, and so there are always males patrolling the forest canopy. On occasion an interloping male flies into an occupied sunlit spot and is challenged by the owner. The two males then make a brief spiral flight up toward the canopy, after which one flies up into the canopy and the other settles back down into

the spot. By marking male butterflies Davies was able to show that it is invariably the original owner that returns to the sunlit spot after a spiral flight.

Once again there are two plausible explanations for such behavior. It is possible that ownership is accepted as a cue and that the spiral flight serves somehow to inform the interloper that the sunlit spot is occupied. Or it is possible that only relatively strong butterflies hold sunlit spots and that the spiral flight serves to demonstrate their strength. Davies favors the first explanation: that the speckled wood butterfly is operating according to the bourgeois principle. He has two reasons for doubting that only strong butterflies hold sunlit spots. To begin with, he noted that most of the males he marked in the canopy were eventually observed holding spots. Moreover, he performed an experiment in which he removed the owner from a

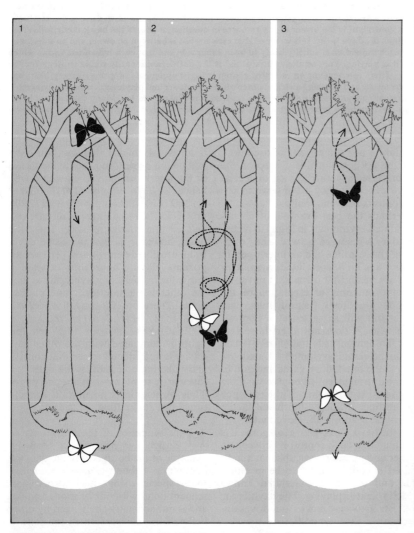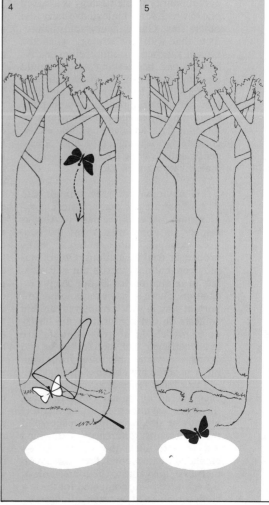

EXAMPLE OF BOURGEOIS STRATEGY has been discovered by N. B. Davies of the University of Oxford. Males of the speckled wood butterfly (*Pararge aegeria*) follow the bourgeois strategy in territorial disputes over sunlit spots on the forest floor. If a male from the forest canopy descends into an occupied sunlit spot (*1*), it is challenged by the owner of the spot. The two butterflies execute a short spiral flight up toward the forest canopy (*2*), after which the original owner returns to the sunlit spot and the interloper returns to the canopy (*3*). There is additional evidence to suggest that ownership is accepted as a conventional cue in this species and that the spiral flight serves somehow

sunlit spot, waited until a new male descended from the canopy and then released the original owner back into the spot. On each occasion the new owner won the ensuing dispute and the original owner retreated.

This last experiment suggests that a male of this species considers that it owns a sunlit spot when it has settled unchallenged in the spot for a few seconds. What happens if two males consider themselves to be owners of the same spot? Davies investigated the question by surreptitiously introducing a second male into occupied territory. Sooner or later one of the owners would notice the other and challenge it. In all cases there ensued a protracted spiral flight lasting an average of 10 times longer than a normal flight. It appears that a butterfly perceiving itself to be an owner is prepared to escalate.

Not all asymmetric contests are as simple as the ones I have described so far. For example, contests between male fiddler crabs of the species *Uca pugilator* can involve an asymmetry between the owner of a burrow and a wandering crab and also asymmetries in the crabs' size and strength. It appears that a large component of the behavior of the crabs is concerned with assessing the asymmetries. Gary W. Hyatt and Michael Salmon of the University of Illinois found that in 403 fights between males of this species the burrow owner prevailed in 349 instances. In the 54 fights won by the wandering male that crab was larger than the burrow owner in 50 instances and smaller in only one. It is evident that asymmetries in ownership and fighting ability are relevant, but it will not be an easy task to develop and test a game-theory model that can help to explain the form and duration of the contests.

There are many relevant factors in animal contests that I have not discussed. For example, in some instances the contested resource is divisible, so that sharing it may be preferable to fighting over it. Some animals provide false information about their size (as with a ruff or a mane) or their intentions. Game-theory models will have to be devised that take account of these features. There are also different types of animal behavior for which game-theory models are appropriate. For example, parental care is not normally regarded as a contest because parents have a common interest in the survival of their offspring. The activity has areas of conflict as well as areas of common interest, however, and I believe a game-theory analysis is illuminating. Finally, in seeking the differences and similarities between man and other animals, it may be helpful to analyze the "games" they can play.

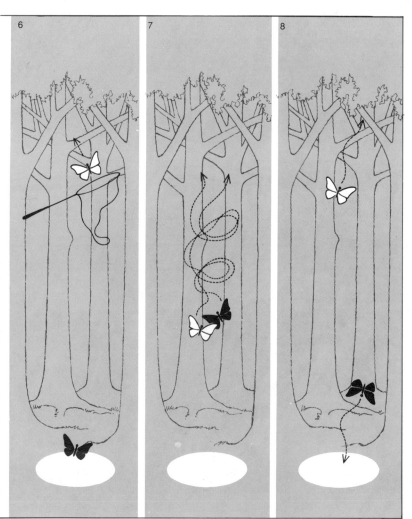 

to inform the interloper that the sunlit spot is occupied. If the owner is removed from a spot (*4*), another male will descend to occupy it (*5*). When the original owner is placed back in the spot (*6*), a spiral flight ensues (*7*), and this time it is the new owner that returns to occupy the spot (*8*). If two male butterflies consider themselves to be owners of a single sunlit spot, one challenges the other and they execute an extremely long spiral flight (*9*) from which either one can emerge the winner (*10*). It appears that, just as the hawk-dove-bourgeois game predicts, a male speckled wood butterfly that considers itself to be the owner of a sunlit spot is willing to escalate a contest.

Social Spiders

3

by J. Wesley Burgess
March 1976

*Most adult spiders lead solitary lives. A few species,
however, are gregarious and others even build large
communal webs. Both degrees of spider sociality can
be observed among species native to Mexico.*

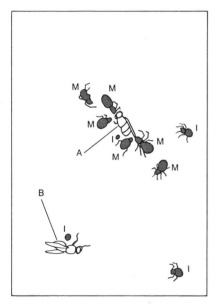

COOPERATIVE CAPTURE of a fly (*A*) by
several spiders is seen in the photograph on
the opposite page. The diagram above identifies prey and predators. Spiders labeled *M*
are mature; those labeled *I* are immature.
Only two of the many flies on the web (*A, and
B at bottom left*) are new catches. The spiders
are of the social species *Mallos gregalis*. The
cluster of mature spiders is feeding or prepar-
ing to feed. One immature spider has been
drawn to the scene; another is approaching.
The photograph was made in the author's
laboratory; spiders were collected in Mexico.

Among insects—notably bees, ants and
termites—social behavior is common. Among spiders it is rare. All
spiders are predatory carnivores; among
many of them even the male of the species
cannot approach the female without risk of
being attacked and killed. It is therefore
paradoxical that there are any social spiders
at all. How, then, can such spiders exist?

The number of social spiders is small;
only in 12 genera distributed among nine
families of spiders is any kind of sociality
known. The 12 genera are, however, widely
distributed, with representatives in both the
Old World and the New. Two of the New
World species are found in Mexico. I re-

cently visited areas near Guadalajara where
both species are present, observed the social
spiders in their natural habitat and brought
home to North Carolina a number of specimens for rearing and further observation in
the laboratory.

The two Mexican species lead distinctive-
ly different lives. *Mallos* (formerly *Coeno-
thele*) *gregalis* is a small spider, with a body
that rarely exceeds five millimeters in
length. It builds a large colonial web, surrounding the branches of a tree with a continuous sheet of silk. Its aggregations may
be socially the most complex spider colonies
in North America. *Oecobius civitas* is an
even smaller spider; few have bodies more
than two and a half millimeters long. It lives
gregariously, spinning its silk shelter and
alarm-system web in a dark and narrow
microhabitat: the underside of a rock.

Spider societies are different from the societies found among the higher social insects both in kind and in degree. One reason
is that a spider's web extends its range of
sensory perception in a way that has no
analogy among insects. Another is that the
structure of a spider's mouthparts is such
that it can feed only on other animal life.
Any animal of appropriate size that a spider
encounters, including a spider of another
species or even the same species, is potential
prey. It will nonetheless be useful in describing the sociality of the social spiders to
sketch the probable evolution of different
degrees of sociality among insects.

As Edward O. Wilson of Harvard University has pointed out, the eusocial insects,
or higher social insects, have three traits in
common: cooperative care of the young, a
division of labor whereby more or less sterile individuals attend to the needs of fertile
individuals, and a life cycle long enough for
the offspring at some point to share the activities of the parental generation. The evolutionary routes that may have led from
nonsocial to eusocial behavior appear to be
traceable in terms of the less than eusocial
behavior found among various insect relatives of the eusocial species. Charles D.
Michener of the University of Kansas has
outlined two such possible routes.

The first route Michener calls parasocial;
on it there are three levels of increasingly
complex behavior on the way to eusociality.

The lowest level, communal behavior, is
characterized by an aggregation of female
insects, all belonging to the same generation; once the females have aggregated they
build a communal nest for their young. The
next level, quasi-social behavior, is characterized by cooperative care of the young.
The third level, semisocial behavior, is characterized by the appearance of different
castes that serve different roles. Thereafter
eusociality is achieved when the life cycle is
extended so that parents and mature offspring coexist in the same colony.

Michener's second evolutionary route he
calls subsocial. On this route only one level
of behavior precedes eusociality; it is characterized by solitary rather than communal
nest building. The solitary female remains
at the nest, however, and cares for her
young. Eusociality is achieved in one step
when the nest builder lives long enough to
have the assistance of its first daughter generation in caring for subsequent, caste-dif-
ferentiated daughter generations.

Looked at in these terms no social spider is
eusocial. We must define the common
base of spider sociality in much more restricted terms: the existence of various degrees of communality and of characteristic
interactions among the members of communal aggregations.

Here it should be noted that with few
exceptions even spiders that are solitary in
habit go through a semicommunal stage
early in their life cycle. Unlike insects,
spiders do not have a larval stage. Each
emerges from the egg as a functioning mini-
ature adult, although it retains a yolk sac
that supplies it with nutrients for several
days. It grows in size and develops its sexual
characteristics through a series of successive molts, the earliest of which takes place
within the shelter of the parental egg sac. It
leaves the egg sac fully prepared to spin silk
and disable prey.

One might therefore expect that the spiderlings of the solitary species would scatter
as soon as they leave the egg sac. Instead for
the duration of a period known as the tolerant phase the spiderlings aggregate, and
many of them join in the labor of building a
small sheet web. They may even attack any
small prey animal that blunders into the

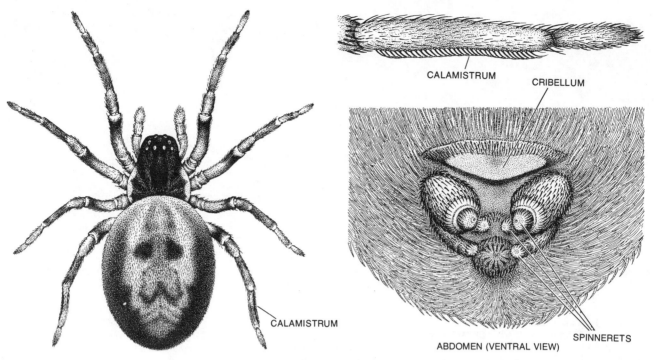

CALAMISTRUM

CRIBELLUM

CALAMISTRUM

SPINNERETS

ABDOMEN (VENTRAL VIEW)

COMMUNAL SPIDER *Mallos gregalis* **has an average body length of five millimeters. Its complex flytrap web incorporates many sticky bands of silk that entangle intruders. The sticky silk is drawn from** hundreds of microscopic pores in an abdominal plate (*right*), the cribellum. The spider combs out the silk with its calamistrum, an array of bristles that grows on the metatarsal of each of its hind legs.

web and wrap the intruder in silk. At this early stage, however, they never feed on the prey. After several days of the tolerant phase have passed the spiderlings disperse, build individual webs and feed on the prey they capture. All the spiderlings appear to adopt the solitary behavioral pattern simultaneously.

It is also noteworthy that in certain solitary-spider species (including representatives of the families Eresidae, Theridiidae and Agelenidae) the adult female does not abandon the egg sac after constructing it but remains with it, or carries it with her, until the spiderlings emerge. The female may then allow them to share her captured prey or may nourish them with regurgitated food or special secretions. Such parental care of the offspring bears a certain resemblance to the lower level of Michener's subsocial route to higher sociality. Thus even among the spider species that are recognized as being solitary, transient episodes of sociality may be observed.

When spiders live in groups, a number of additional interaction patterns are evident. Spider groups form in a variety of ways. For example, adult spiders of some species in the families Uloboridae and Araneidae will aggregate without regard to whether they are the offspring of the same parents or different ones. Each individual in these aggregations spins its own web. Among some species the individual may also contribute silk to a communal web area. Some of these groups may be made up of as many as 1,000 adults. In general each individual lives independently. All, however, share the benefits of a large aggregate web surface and of monopolizing a habitat that might otherwise

have been shared by competitive species.

The viability of simple aggregations such as these demonstrates the existence of a tolerance mechanism in the individual adult spiders. At the very least the mechanism must be strong enough to keep the spiders from eating one another when prey are scarce. Evidently the mechanism is also species-specific; it is not limited to simply ensuring that the spiders are tolerant of all the other spiders in the aggregation. They are also tolerant of any spider of their own species. This has been demonstrated as follows. Individuals of the species *Metepeira spinipes,* a member of the family Araneidae, have been taken from populations living hundreds of miles away and introduced into local aggregations of *M. spinipes.* The presence of strangers did not disrupt the tolerance mechanism within the local aggregation, nor was any difference noted in the behavior of the two groups.

The most dramatic examples of spider sociality involve interactions substantially more complex than those I have been describing so far. These interactions are known only for four (possibly five) spider species. Two of the species are African: *Agelena consociata* and *Stegodyphus sarsinorum.* The others are New World spiders: *Anelosimus eximus* (and possibly a second species of the genus, *A. studiosus*) in South America and one of the species I have collected in Mexico, *Mallos gregalis.* All have in common the habit of constructing a large central web that is occupied by all the spiders in the aggregation. By combining their labors the spiders are able to construct a web that is much larger and far more elabo-

rate in architecture than the web of any single spider; the structure is occupied by successive generations.

These spiders also collaborate in capturing prey much larger than prey any one of them could capture alone. Moreover, after the prey has been captured the spiders feed on it communally. Interactions as complex as these imply that these species have in addition to a tolerance mechanism a capacity for the coordination of individual responses to stimuli and an ability to recognize intraspecies sensory cues or to respond to some other kind of information. As an example, each spider seems to be able to distinguish between the web vibrations caused by a fellow member of the community and the vibrations caused by potential prey.

Bertrand Krafft of the University of Nancy has observed *Agelena consociata* in Gabon. He found that close-quarters tolerance in the species is mediated at least in part by chemotactic cues. Uninjured members of the community tolerate one another. An injured spider, or one whose normal superficial odor has been artificially altered by a washing in alcohol and ether, is attacked immediately. Neither the chemotactic cues nor other possible but still unidentified components of the spiders' tolerance mechanism are confined to local populations of the species. As with *Metepeira spinipes,* individual spiders of the same species can be moved from one colony to another without disrupting the communal activity pattern.

There is no evidence that any of these spider species has evolved a caste system such that the adults differ in form in accordance with any division of labor. Some difference in behavioral roles may exist as a result

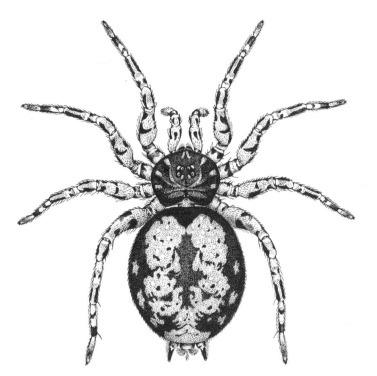

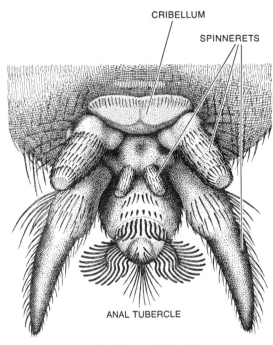

CRIBELLUM

SPINNERETS

ANAL TUBERCLE

ABDOMEN (VENTRAL VIEW)

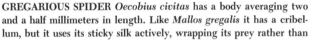

GREGARIOUS SPIDER *Oecobius civitas* has a body averaging two and a half millimeters in length. Like *Mallos gregalis* it has a cribellum, but it uses its sticky silk actively, wrapping its prey rather than waiting for it to become entangled. The spider combs out a thread of the silk with its anal tubercle (*right*) and winds silk around the prey by circling it, abdomen foremost (*see middle illustration on page 34*).

of age or variations in biological rhythms, but just how cooperation is cued remains unknown. The pattern of behavior is nonetheless an example of sociality that is not easily equated with any pattern of sociality among insects. These spiders' behavior may well deserve a category of its own: communal-cooperative.

The Mexican social spider *Mallos gregalis* traps mostly flies on the sticky surface of the communal sheet web it spins around the branch of a tree. The spider has long been known to Mexicans as *el mosquero,* the fly-killer, and in the rainy season, when houseflies are particularly oppressive, those who live in the Guadalajara countryside will bring a web-covered branch into their house in much the same way that other people might string up flypaper. A member of the family Dictynidae, *M. gregalis* is a cribellate spider. Such spiders have a sievelike plate, the cribellum, on the underside of their abdomen [*see illustration on opposite page*]. Sticky silk emerges from fine holes in the cribellum and is combed away with the two hind legs that bear a special row of bristles known as the calamistrum. This is the silk that forms the sticky prey-trapping areas on the outside of the spider's web. The web as a whole is an elaborate structure that includes supporting lines running between the surface sheet and the twigs and leaves of the branch, sheltered retreats for the spiders and special chambers where the female spiders live with their egg sacs. The sacs, thin wrappers of silk, contain from 10 to 20 eggs. The surface sheet is perforated in places with holes that provide access to the interior of the web.

The communal web of *M. gregalis* can be very large. One I saw near Guadalajara covered the limbs and branches in the upper three-quarters of a 60-foot tree of the mimosa family. Where the limbs met the trunk the silk of the sheet web was gray, but near the tips of the branches the silk was new and white. Evidently construction was continuing outward along the limbs. The spiders were not confined to the newer portions but were active in all parts of the web.

Both field and laboratory observations confirm that the construction of the *M. gregalis* web is a mutual effort. If a laboratory colony of the spiders has some treelike support available, such as an upright stick, the spiders will build their characteristic enveloping sheet web. In the absence of such a support they will build the kind of three-dimensional web that is typical of other dictynid species. Although this web looks different from the natural one, it too includes retreats and egg-sac chambers. In the laboratory web a task begun by one spider may be finished by another. I have also seen one spider of the colony lay down strands of ordinary silk, after which other spiders added bands of the sticky cribellate silk.

Observed in nature, the spiders seem to move around at random and without haste, emerging from and disappearing into the holes in the surface of the web. Their fly prey are trapped by the sticky web when they alight on it. When a fly gets stuck, two or three spiders approach the buzzing insect, immobilize it with their venomous bites and then feed on it. On occasion the spiders can be seen carrying flies down the holes into the interior of the web.

The spiders' predatory behavior can be observed in detail in the laboratory. We feed our colonies once every five days, which increases the probability that a majority of the spiders will be ready to feed at the same time. At any given moment one or two individuals in a colony of about 100 spiders are usually on the surface of the web; the other spiders will be in the web's interior. When a housefly is put into the spider cage and flies about, it makes a humming noise that is audible to the experimenter but causes no apparent change in the random activity of the spiders.

A fly that lands on a nonsticky part of the web and walks around stimulates a localized response; some of the spiders will turn to face in the direction of the fly, but that is all. If the fly gets entangled in a sticky part of the web and begins to buzz loudly, the behavior of the spiders changes abruptly. Throughout the web spiders that have been at rest turn toward the trapped fly and begin to approach it in short jumps. The fly continues to buzz even after the first spiders to reach it start their attack, usually by biting a leg or a wing. The buzzing draws more attackers; they move directly toward the fly over the web surface until eventually the prey almost disappears under the feeding spiders. Both male and female spiders attack. Even immature spiders take part, swarming over the adults in search of a place to feed.

Even though the attacking spiders in the caged colony are quite aggressive, we have never observed one spider attacking another. As we canvass the behavioral repertoire that differentiates social spiders from solitary ones, this aspect of feeding in aggregations is significant. For example, young soli-

RESPONSE TO AN INTRUDER by a colony of communal spiders is reconstructed in these drawings on the basis of laboratory observation. Unlike the members of a wild colony all the spiders in the laboratory colony have fasted the same length of time. The sound of a passing fly is audible to a human observer but attracts no attention from the spiders. Even when a fly lands on the surface of the web (*top*

left), only the nearby spiders reorient themselves. The buzzing of a fly entangled in sticky silk (*top right*) stimulates a response throughout the colony, and the spiders advance on the prey in quick jumps. The bites of the first spiders to reach the fly (*bottom left*) give rise to a louder buzzing that further stimulates spiders to approach the prey. As feeding begins (*bottom right*) immature spiders join the adults.

tary spiders such as those of the species *Araneus diadematus,* when they are artificially confined in close quarters, will also feed communally. Among the artificially confined solitary spiderlings, however, a tolerance mechanism, if it exists at all, operates only imperfectly; they will feed communally both on captured flies and on one another. This suggests that it is a strong tolerance mechanism that accounts for communal feeding in *Mallos gregalis* just as coordination mechanisms account for communal capture of prey.

The tolerance mechanism at work in *M. gregalis* colonies is being studied in our laboratory. It is evident from our observations that the mechanism is strong and that it operates both at close quarters and over considerable distances. Indeed, several separate mechanisms may be at work, perhaps mediated by cuing systems that allow discrimination between, say, the web vibrations caused by trapped prey and those caused by members of the colony. To test this hypothesis we are subjecting the colonies to the stimuli of various web vibrations in the hope of isolating such cues.

The social behavior of the second Mexican spider, *Oecobius civitas,* at first seems to be principally aggregative, like the behavior of other spiders that build their nests in close proximity. The darkness of this spider's microhabitat makes observation of its behavior difficult, but its unusual method of prey capture has been recorded. *O. civitas* has a finger-shaped organ, the anal tubercle, on the abdomen near its silk-extruding spinnerets. With this appendage it can comb sticky silk out of its cribellum in a rope that it winds around its prey [*see illustration on page 31*].

A closer study of the sociality of *O. civitas* proves that it is more than merely aggregative. The spiders' behavior features a curious combination of tolerance and avoidance. On the underside of the rock that shelters the spiders each individual weaves a small open-ended tube of silk that is its hiding place; around this retreat the spider constructs a thin, encircling alarm-system net close to the surface of the rock. The pair of structures makes up the spider's web, which is generally found in a hollow or a crevice of the rock. If a spider is disturbed and driven out of its retreat, it darts across the rock and, in the absence of a vacant crevice to hide in, may seek refuge in the hiding place of another spider of the same species. If the other spider is in residence when the intruder enters, it does not attack but darts out and seeks a new refuge of its own. Thus once the first spider is disturbed the process of sequential displacement from web to web may continue for several seconds, often causing a majority of the spiders in the aggregation to shift from their home refuge to an alien one.

Field observations and experiments indicate that, as with *Metepeira* and *Mallos,* the mechanisms responsible for the combination of tolerance and avoidance extend be-

WEB-COVERED BRANCHES of a species of mimosa near Guadalajara support part of the communal web of a *Mallos gregalis* colony. Scattered holes allow the spiders to move freely from areas inside the web to the sticky outer surface where intruders become trapped. In the fly season local people often bring such branches indoors to serve as a kind of natural flypaper.

yond the local population to include other spiders of the same species. Moreover, within the local population the shift to another spider's shelter may be a semipermanent move. The reason is that when the spiders are undisturbed, they occupy a fixed web position for long periods. In any event the behavioral pattern of the species benefits the individual spider by providing more than one available retreat in an emergency.

The group behavior of *Oecobius civitas* is far simpler than that of *Mallos gregalis.* It is nonetheless effective in enabling the spiders to live together under crowded conditions. No doubt the avoidance mechanism makes a major contribution toward the spiders' ability to maintain a high population density in their restricted microhabitat. Other contributing factors probably include the spider's unusual predatory technique and the spacing of individual webs. In any case, although we remain largely ignorant of the mechanisms underlying avoidance and tolerance, they appear to be the basic building blocks that provide a foundation for more complex group behavior.

It has been suggested that *Oecobius civitas* displays an even more remarkable kind of sociality: construction of a communal egg sac by the females in the aggregation. The possibility of such a behavioral advance, unknown among spiders, came to light recently when William A. Shear of Hampton-Sydney College undertook a tax-

onomic review of the oecobiid spiders. He was assisted by a number of colleagues who donated specimens to the project. Among the donors was Willis J. Gertsch, curator emeritus of spiders at the American Museum of Natural History, who had collected specimens of *O. civitas,* its web and its egg sacs in the Guadalajara area.

The usual oecobiid egg sac contains from five to 10 eggs. In the preserved material donated by Gertsch, however, Shear found two groups of more than 200 immature spiders. Each group was contained in what gave every appearance of being a single egg sac. Shear published his observation in 1970, suggesting that *O. civitas* might be a communal egg layer.

When I collected specimens of *O. civitas* and its egg sacs in the area near the shores of Lake Sayula, where Gertsch had done his collecting, I found that several other species of spiders shared the rocky habitat with the oecobiids. As a result a variety of egg sacs could be collected. This I did, sealing individual egg sacs in individual tubes. I was disappointed to find that only the small sacs, averaging seven eggs to a sac and mainly collected in or near *O. civitas* web retreats, hatched oecobiids.

After rearing this spider in the laboratory for three generations and observing only individual egg sacs containing from five to 10 eggs, I consider that to be the normal pattern of reproductive behavior in *O. ci-*

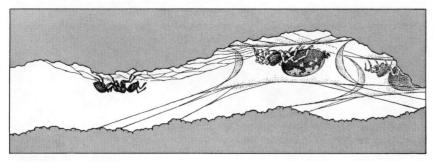

CAPTURE OF PREY, usually a foraging ant, by a spider of the gregarious species *Oecobius civitas* follows a complex pattern that begins when the intruder disturbs the spider's alarm web.

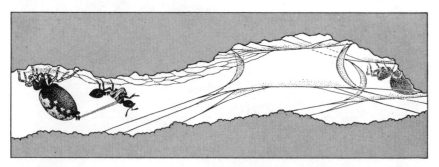

ALARMED SPIDER leaves its shelter and moves in circles around its prey, its abdomen foremost and raised clear of its legs, while it combs out a strand of sticky silk with anal tubercle.

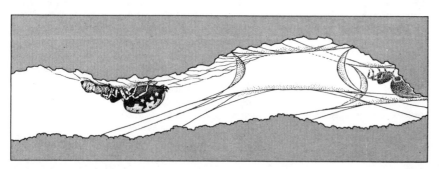

WRAPPED IN SILK, the ant is immobilized. The spider may rest for a time or may turn (*left*) to bite and disable its prey. Only the captor feeds on the prey; nearby spiders do not approach.

vitas. To resolve the question beyond all doubt other single *O. civitas* egg sacs containing eggs or immature spiders in large numbers will have to be collected in the field.

Mating behavior has not yet been observed in our laboratory populations of either *Mallos gregalis* or *Oecobius civitas*. Solitary male spiders go through elaborate pre-mating maneuvers, so-called courtship patterns that supposedly inhibit predation in the female at the time of copulation. Among social spiders, which live in tolerant aggregations, such maneuvers would not seem necessary. Indeed, if differences in copulatory patterns between solitary and social spiders do exist, they may even provide clues to the evolutionary background of spider sociality. In this connection we have made one possibly significant observation concerning fertility. Solitary spiders raised in the laboratory retain the cyclical breeding rhythms characteristic of their wild state, but when our *M. gregalis* colonies are provided with a uniform environment and controlled periods of darkness and light, they produce fertile eggs throughout the year.

Observation of the two Mexican spiders has uncovered a substantial amount of information about their sociality, but that information more often than not merely defines the extent of our ignorance. For example, we do not know what conditions favor the development of spider sociality or even what mechanisms are involved in tolerance, avoidance, the formation of groups or the coordination of activity. Moreover, it is not known how different forms of spider sociality are related to one another or how, in complex interactions, intragroup information is transferred. The search for answers nonetheless seems to offer one certainty: The more we learn about the sociality of comparatively simple animals, the better we shall be able to understand the sociality of more complex species, including our own.

The Social Behavior
of Army Ants

4

by Howard R. Topoff
November 1972

*The complex and permanent social organization of
these insects is maintained by interactions among
a great many individuals. Each individual, however,
can alter its behavior only slightly.*

Every living organism must at some time interact with other members of its species. As a result some degree of social behavior is the rule in the animal kingdom. The late T. C. Schneirla, for many years a curator in the department of animal behavior of the American Museum of Natural History, was greatly interested in the evolution and development of social behavior in species of animals representing many levels of evolutionary history. One of the groups of animals he selected for study was the army ants. These ants were already famous for their awesome marches in masses of hundreds of thousands of individuals. Their notoriety had given rise to an abundance of military metaphors, exemplified by the following description by A. Hyatt Verrill, a naturalist and explorer of animal life in South America: "In all the world, the army ants of the tropics are the most remarkable in many ways. Utterly blind, yet they move in vast armies across the land, overcoming every obstacle other than fire and water, maintaining perfect formation, moving with military precision and like a real army having their scouts, their engineering corps and their fighting soldiers."

In 1932 Schneirla proceeded to study army ants more objectively, and he discovered by close and long-continued observation of their habits that colonies of army ants show a degree of social organization every bit as impressive as the apocryphal stories of the early naturalists [see "The Army Ant," by T. C. Schneirla and Gerard Piel; SCIENTIFIC AMERICAN, June, 1948]. How were the ants' activities organized and coordinated? Schneirla's studies gave rise to various hypotheses, and the nature of the social bond in these insects continues to intrigue investigators of animal behavior.

During the past 10 years my colleagues and I have been continuing these investigations in an attempt to better understand the complex mechanisms that organize the remarkable social behavior of the army ants.

The army ants comprise one of the subfamilies of ants known as Dorylinae (so named from the Greek word meaning spear, because of their potent sting). There are some 150 species of army ants in the Western Hemisphere, most of them in Latin America, and about 100 other known species in Africa, Asia, Indo-Malaysia and Australia. The majority of species inhabit tropical regions, but many have adapted to temperate climates; some 20 species are found in the southern and mid-central U.S.

Practically all army ants are notable for four characteristics. The first is that they typically have very large colonies. Even small colonies, such as those of the Asian genus *Aenictus*, consist of at least 100,000 individuals. *Eciton burchelli* of Central and South America has colony sizes approaching a million individuals. In certain species of the African genus *Dorylus* colonies have been estimated to run to more than 20 million individuals. The second characteristic is the ants' periodic shifting of nesting sites. These colony movements, or emigrations, involve the entire colony: workers, brood and queen. At the end of each emigration the army ants settle down only in temporary bivouacs.

The third characteristic is that the army ants are almost exclusively carnivorous. They feed on other arthropods, particularly insects, and some species have been observed to eat small vertebrates such as lizards and snakes. The raiding parties of certain species are so huge that wherever they live the ants

rank as major predators in the ecological community. Edward Step, an observer of insect life, described the army ants' forays in these terms: "They march in such enormous numbers that everything which desires not to be eaten has to fly before them; from the cockroach to the mouse to the huge python, the elephant, the gorilla and the warlike native man, the story is the same." This is a gross exaggeration, but it is a well-authenticated fact that the ants' food consumption is tremendous; the workers of a colony of *Eciton burchelli*, for example, may bring in more than 100,000 other arthropods a day to feed the nest.

The last characteristic that typifies the behavior of doryline ants is their tight cohesiveness. Unlike many species of ants from other subfamilies, army ants do not forage for food individually. Instead all raiding and emigrations to new bivouacs are conducted by groups of individuals that closely follow a chemical trail deposited continuously by all the ants as they run along the ground.

To find an explanation of the social behavior of these animals we must look into their physiology and their means of communication, which, since they are essentially blind, is based mainly on chemical and tactual stimuli. Let us first examine the army ants' life-style. We shall consider specifically three species that are well known because they conduct their activities principally aboveground: *Eciton hamatum* and *Eciton burchelli* of Central and South America and *Neivamyrmex nigrescens*, which ranges into temperate climates and is found in the U.S. The three species are closely related evolutionarily and are much alike in their life cycle and behavior.

Colonies of army ants typically con-

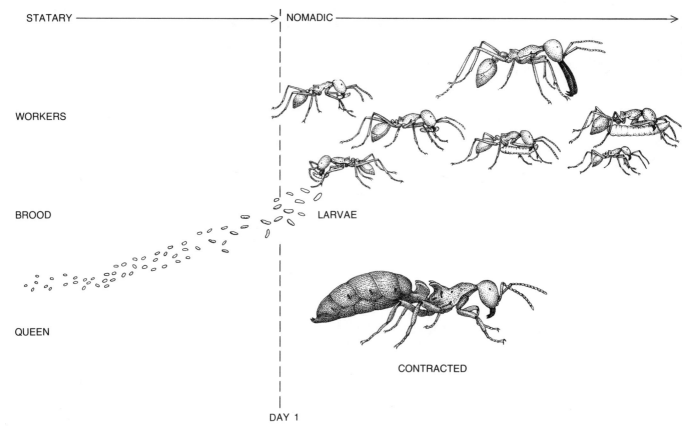

ALTERNATING PHASES in the behavior of army ants of the New World tropical genus *Eciton* are illustrated here, beginning with Day 1 of a nomadic phase (*left*). For 14 to 17 days after a new generation of workers emerges from the pupal stage a colony of army ants sends out large parties of raiders for food every day. Every night the colony shifts to a new bivouac, carrying the larvae that will become the next generation of workers. During the nomadic phase the queen's abdomen remains contracted. The phase

sist of a single queen, a brood of developing young and a large population of adult workers. The queen is the colony's sole agent of reproduction, and she is responsible to a great extent for the colony's cohesion. The queen secretes certain substances that are attractive to the workers and therefore hold the colony together. More important, the chemical secretions of the queen actually enhance the survival of the workers (as has been shown by Julian F. Watkins II of Baylor University and Carl W. Rettenmeyer of the University of Connecticut). At regular intervals (about every five weeks) the queen's large abdomen swells with fatty tissue and eggs, and she may lay well over 100,000 eggs in the course of a week. The eggs then give rise to four successive stages of development: embryo, larva, pupa and adult, which on emergence from the pupal stage is lightly pigmented and readily recognizable as a "callow," or young worker. The workers are all female but sterile, with underdeveloped ovaries.

In species of *Eciton* and *Neivamyrmex* the workers developing from a given batch of eggs vary in size and structure,

a condition known as polymorphism. The developmental basis for polymorphism is not clear, although two possibilities exist. The first is that the developmental pathways leading to adult workers that differ in size and structure are determined by the amount and kind of food eaten by the larvae immediately after they hatch from the egg. This mode of development exists in honeybees. The second possibility is that the eggs laid by the queen are so different biochemically that the subsequent development of the larvae is unaltered by the quantity or quality of food consumed. (In some genera, such as *Aenictus,* the adults are not polymorphic.) Whatever the differentiating mechanisms may be, the adult ants that differ in size and structure also exhibit contrasting patterns of behavior, with the result that there is a division of labor in the colony. Small workers (as little as three millimeters in length) spend most of their time in the nest feeding the larval broods; intermediate-sized workers constitute most of the population, going out on raids as well as doing other jobs. The largest workers (more than 14 millimeters in species of *Eciton*) have a huge head

and long, powerful jaws. These individuals are what Verrill called soldiers; they carry no food but customarily run along the flanks of the raiding and emigration columns. An excited "soldier" is a formidable animal: it rears up on its hind legs, vibrates its antennae and rhythmically opens and closes its jaws. The tips of the mandibles are extremely sharp and are curved backward; if the ant bites a human being, they penetrate the skin and are difficult to remove.

In the colony's behavioral cycle there is a "nomadic" phase during which a large proportion of the adult workers go out on daily raids and collect food. In both species of *Eciton* the raids begin at dawn. The ants pour out of the bivouac and form several columns, each column later dividing into a network of branches. In running along these trails the ants seem not to depend much on vision. Species of *Eciton* and *Neivamyrmex* have vestigial eyes consisting of only a single facet; they can detect changes in the intensity of light but almost certainly cannot reproduce an image of an object. The raiding ants stay together by following a chemical trail laid on the ground by the other workers.

STATARY ————————————————————————————————→| NOMADIC —————————————→

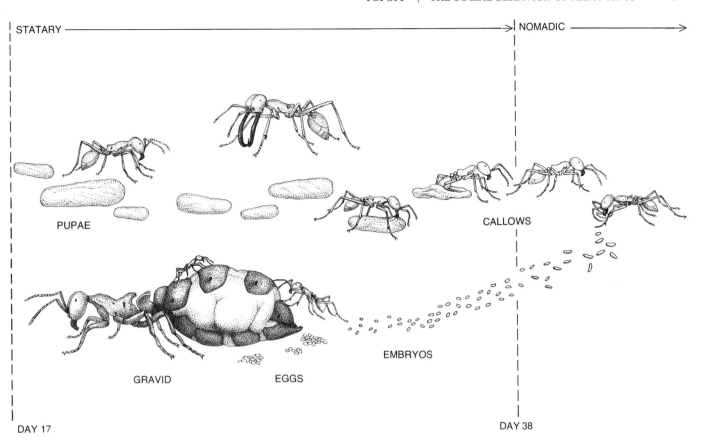

PUPAE

CALLOWS

GRAVID   EGGS   EMBRYOS

DAY 17   DAY 38

ends when the colony's larvae enter the pupal stage. The statary, or resting, phase then begins (*center*). The colony ceases to move nightly and the daily raiding parties are smaller. The statary phase continues for some 21 days. During the first week the queen's ab-domen enlarges rapidly. During the second week she may produce as many as 100,000 eggs. By the end of the third week, as a new generation of worker ants completes the pupal stage, the queen's eggs hatch into larvae. This initiates a new nomadic phase (*right*).

Murray S. Blum of the University of Georgia and Watkins have determined that the substance deposited by the army ants originates in the hind intestine, but it is not yet established whether the substance consists of undigested food or a glandular secretion or a combination of both.

By midmorning the raiding columns have overrun an area extending a considerable distance from the nest, often more than 100 meters. At the front of the raiding columns the ants attack insects and other arthropods, biting and stinging the prey, pulling it apart and carrying the softer pieces back to the nest. Thus the column is actually a two-way stream, with some ants advancing and others returning with their prey.

At nightfall the entire colony moves on to a new bivouac, typically emigrating along one of the principal raiding trails of that day. It may take the colony most of the night to complete the move to the new nesting site. This daily routine of massive raids and emigrations from bivouac to bivouac is followed for 14 to 17 days. Then, more or less abruptly, the colony settles down to a much quieter phase. Comparatively few of the

workers go out on raids; their forays are much smaller and the colony stops emigrating and remains at the same nesting site. This "statary" phase lasts approximately three weeks. At the end of that time the cycle begins again; the colony resumes rushing out on great daily raids and making nightly emigrations.

The foregoing pattern is typical of the *Eciton* ants' behavior. The cycle in species of *Neivamyrmex* follows a similar pattern but differs in some aspects, depending on differences in the habitat; for example, in an area of high daytime temperature and low humidity the ants conduct both their raiding for food and their emigrations at night instead of during the day.

In 1932 Schneirla set out to learn what factors regulate the cycles of behavior in *Eciton hamatum* and *Eciton burchelli*. At the time there were already two rival hypotheses. One suggested that the cycles of behavior were influenced by physical conditions of the environment, such as temperature, humidity, air pressure or the phases of the moon; the other suggested that the stimulus for emigrations might simply be depletion of the food supply in the area around the biv-

ouac. Schneirla soon showed that both of these conjectures must be incorrect. He found that generally within a given environment some of the colonies were in the nomadic phase and some were in the statary phase; that ruled out environmental conditions as the determinant of whether or not a colony would make nightly emigrations. Schneirla disposed of the second hypothesis by observing that a colony of army ants would sometimes move into a nesting site that had just been vacated by another colony, and the newcomers would remain at this site even for a three-week statary period—clear evidence that the food supply around the bivouac had not been exhausted.

The actual regulator of the ants' nomadic and statary behavior, as Schneirla eventually demonstrated, was not some external influence but the breeding cycle within the colony. He noted that the nomadic phase always coincided with the period when a larval brood was developing in the colony, and that the statary phase began when the larvae started to spin cocoons and went into the pupal stage of development.

When army ants emerge from the

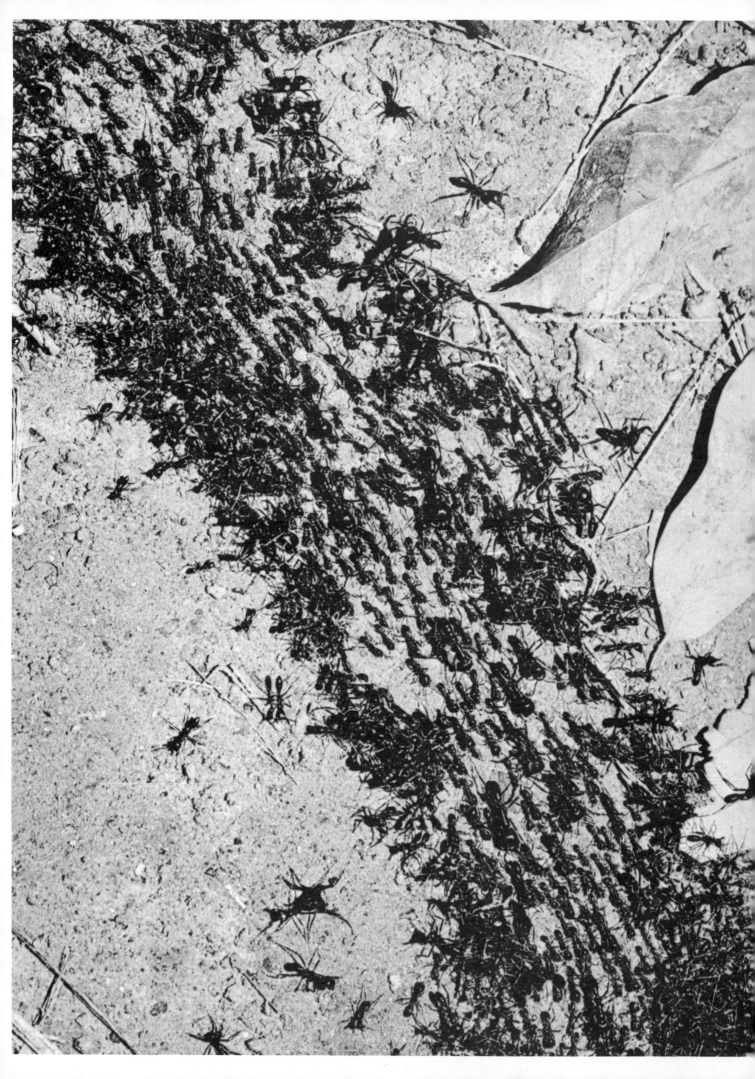

MARAUDING COLUMN of army ants forms the diagonal ribbon in the photograph shown at left. These are ants of the genus *Dorylus*, native to Africa and famous for colonies that may number as many as 20 million ants. The photograph was made by William H. Gotwald, Jr., of Utica College.

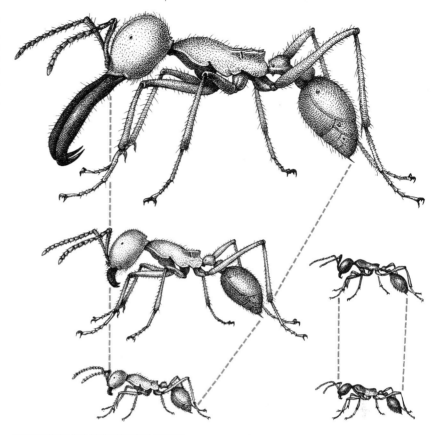

WORKERS OF DIFFERENT SIZES are common in some army ant species and virtually absent in others. Illustrated here are workers of the species *Eciton burchelli* (*left*) and of the Philippine species *Aenictus gracilis* (*right*), enlarged some seven diameters. Workers that differ in size also differ in patterns of behavior. The smallest *Eciton burchelli* workers do little more than feed and maintain larvae; the largest are the "soldiers" of the colony.

pupal cocoons as young workers, the nest is suddenly stirred to a high level of activity. The important stimuli for this excitation are probably substances secreted by these young callows; the older workers respond to the callows by stroking and licking them and by dropping pieces of food on them. This intense social stimulation, which originates with the interactions between the callows and the older workers, is subsequently transmitted throughout the bivouac by communication among the older adults. The result of this high level of mutual stimulation is that the nomadic phase of massive daily raids and emigrations from one bivouac site to another begins. As the callows mature, their chemical excitatory effects wear off but nomadic activities continue in the colony. These activities are maintained by comparable chemical and tactual stimulation imparted to the adult ants by a brood of developing larvae, which by this time have hatched from eggs laid by the queen during the previous statary phase. When the larvae have completed their development and progressed to the cocoon-wrapped pupal stage, the intensity of the mutual stimulation between them and the adult ants decreases abruptly. At this time the colony again lapses into the statary condition, which continues until the pupal brood once again emerges as callow workers.

Although the precise nature of the chemical interactions between adult workers and the brood is not yet known, the evidence supporting Schneirla's hypothesis is strong. For example, when he removed a larval brood from a colony during the nomadic phase, the colony stopped emigrating and the intensity of its daily raids diminished. By the same token, when he split a colony into two parts, only one of which contained the larval brood, the workers in the portion with the larvae continued to show considerable activity whereas those in the broodless portion became less active.

Schneirla's hypothesis predicts that the large differences in behavior exhibited by colonies of army ants during the nomadic and statary phases must result

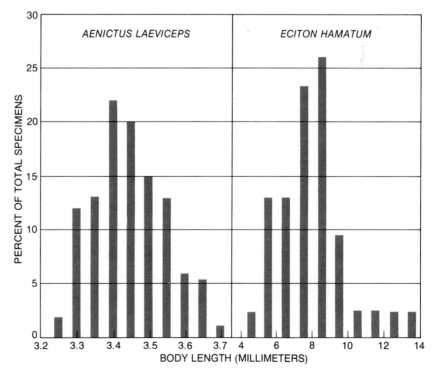

RANGES OF SIZE within two species of army ants are compared in this graph. In *Aenictus laeviceps* the body of the largest worker is scarcely half a millimeter longer than that of the smallest one. In *Eciton hamatum*, however, the body-length difference is nine millimeters.

from corresponding differences in the physiological condition and behavior of each individual ant during the two phases.' At the present time we are a long way from understanding the changes that take place in each ant's endocrine secretory activity, metabolic processes and sensitivity to physical and chemical stimuli. Nevertheless, several experiments have given us considerable insight into the kinds of factors involved.

Many of my own studies have been focused on the underground-nesting, nocturnal army ant species *Neivamyrmex nigrescens,* which I have observed in the field and in laboratory experiments at the Southwestern Research Station of the American Museum of Natural History in Arizona. I noticed that during the nomadic phase the ants not only spent most of the night raiding outside the nest but also frequently set out on their raids late in the afternoon when there was still considerable light on the ground. During the statary phase, in contrast, a smaller number of ants carried out weak and short raids, and they rarely emerged from the nest before dark. How could one account for this apparently slight but significant difference in behavior? One clue comes from studies of the relation between physiology and behavior in other species of animals. As an example, J. Goldsmid of Rhodes University in South Africa described a series of interesting behavioral changes in larvae of the blue tick *Boophilus decoloratus.* For a few days after hatching these larvae are strongly repelled by light and strongly attracted to one another by their common chemical secretions; the mutual attraction is so strong that the larvae come together even in an area under bright light. A week after the larvae have hatched, however, their behavior changes dramatically; they stop responding to one anothers' chemical secretions and do not withdraw from the light.

I was struck by the thought that the ticks' pattern of changing behavior might fit in with my observations of the army ants during the two behavioral phases. Suppose that during the statary phase the workers' negative response to light increases and at the same time their positive response to one another (as well as to the queen and brood) increases. This could account for the fact that during the colony's statary phase workers spend a greater amount of time inside the nest and the raids do not begin until dark. Conversely, if one supposes that during the nomadic phase the ants are less attracted to other individu-

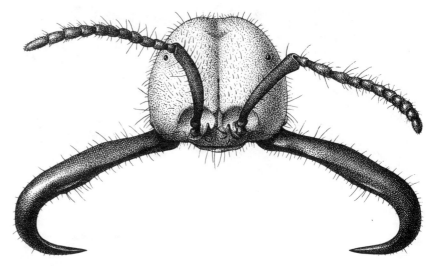

LARGE HEAD AND HUGE JAWS are characteristic of the soldiers in the genus *Eciton*. Once the sharp jaws have pierced an object their hooked tips make them difficult to remove.

als in the nest and less repelled by light, that might explain why most of them readily leave the nest for raiding and why they are not so deterred from starting their raids in daylight.

I designed a series of experiments to test whether or not these ants did indeed shift in their responses to light and to stimuli from other ants during the nomadic and statary phases. The experiments consisted in collecting ants in the field and placing them in an arena in the laboratory where they were able to enter into dimly lighted or brightly lighted areas. In order to minimize handling and artificial excitation of the ants I constructed a cylindrical cartridge in which they were picked up by suction. Without further disturbance of the insects the cartridge was then placed in the center of the arena, where through slits in the base of the cartridge the ants could move into dimly lighted quadrants or into quadrants where the light was 100 times more intense. To monitor the ants' movements we photographed their positions at five-second intervals for two minutes.

The results of this experiment showed that ants taken from colonies during the nomadic phase indeed behaved very differently from those taken from the colony during the statary phase. The nomadic ants traveled about indiscriminately in the bright and dim chambers of the arena, and they ran rapidly in columns with the individuals well spaced out. In contrast, ants in the statary-phase condition either entered into one of the dim areas immediately after the start of the test or else they tended to associate into tight clusters

near the edge of the brightly lighted central cartridge (as the newly hatched tick larvae did in Goldsmid's experiment). When these experiments were repeated with all the experimental areas kept totally dark, the results were just about the same. The record of the ants' movements (filmed by infrared photography) showed that the nomadic ants ran about freely and the statary ants again clustered tightly near the edge of the central cartridge. This supported the conclusion that the ants' attraction to one another changes significantly from the nomadic phase to the statary phase. Although the behavior of the ants in the field and in the laboratory still suggests that they also respond differently to light during the two phases, in the experimental tests their increased attraction to one another during the statary phase overrides their increased negative response to light.

Obviously these experiments are only a beginning in exploration of the interactions that take place among the individuals in an army ant colony during changes in the social behavior of the colony as a whole. Furthermore, we are still a long way from understanding the biological bases of these differences. For instance, does the excitatory stimulation that the adult workers receive from the callows and the larval brood influence their behavior through neural mechanisms alone, or does it also affect the secretory activity of their endocrine glands? Because every adult army ant continuously alternates from the nomadic phase to the statary phase and back again, these ants are excellent species for future studies of the relation between changes in physiological processes and corresponding changes in behavior.

I want to turn now to another interesting question we have been investigating. In primate societies, particularly those of humans, it is well known that newborn individuals do not become fully participating members of the society until they have matured and gained much experience within the family group, with their peers and with other members of the society. Many people believe animals such as insects emerge from the pupal stage of development with an immediate capacity to behave exactly like mature individuals. That is simply not so. With experimental procedures I devised in collaboration with Katherine Lawson, a graduate student at the City University of New York, we compared the behavior of callows and fully mature ants of the genus *Eciton* at the Smithsonian Institution's research station on Barro Colorado Island in the Panama Canal Zone.

The behavior of the callow members of a colony exhibits a puzzling inconsistency. During the first few days after the callows have emerged from their cocoons they do not join the mature adults in predatory raids. Furthermore, if a group of callows is taken from the nest and placed in the midst of a raiding column, they move only sluggishly and in a somewhat disoriented fashion, so that they interfere with the two-way traffic of the rapidly running mature ants in the columns. Yet surprisingly the callows have no hesitancy about going along with the entire colony in the emigration to a new bivouac after the day's raid.

Was the callows' failure to participate in raiding due to an inability to follow the trail deposited by the raiding ants? During the day's raid hundreds of thousands of foraging ants continually run from the bivouac to the raiding areas and then back again. As a result at the end of the raid the strength of the trail may be considerably higher than it is during the early hours of morning. That is a reasonable assumption on the basis of experiments conducted by Richard Torgerson of Wartburg College and Roger D. Akre of Washington State University, who demonstrated that the chemical trails of *Eciton* persist on the ground in the field for at least a week. Perhaps by the time a colony of army ants emigrates the trail is so strong that even the callows are able to follow it. To test this hypothesis we measured the comparative ability of callows and ma-

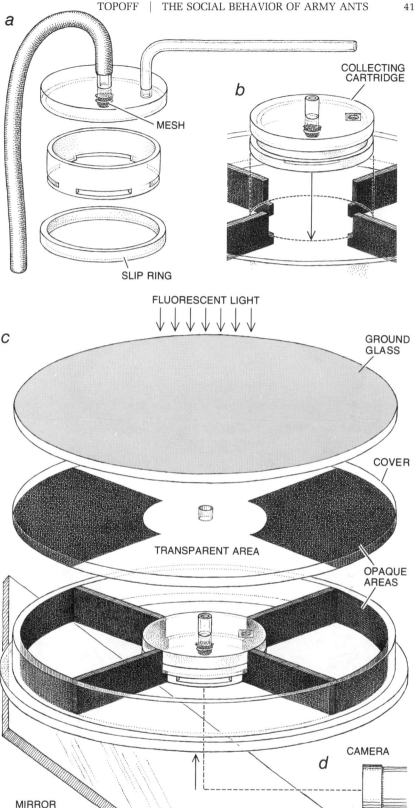

PLEXIGLASS ARENA (c) was constructed to study activity patterns of army ants in the nomadic and statary phases. *Neivamyrmex nigrescens*, a widely distributed New World species, was selected for the study. Ants from nomadic and statary colonies were drawn by suction, some 50 at a time, into collection cartridges (a). They then remained undisturbed until each cartridge was placed in the center of the arena; this raised the ring seal of the cartridge so that the ants were free to explore the arena (b). At first, to assess the ants' response to light, two of the arena quadrants were kept shaded while the other two and the central cartridge were lighted 100 times more brightly. Photographs made at five-second intervals (d) recorded changes in the ants' positions (*see illustrations on following page*).

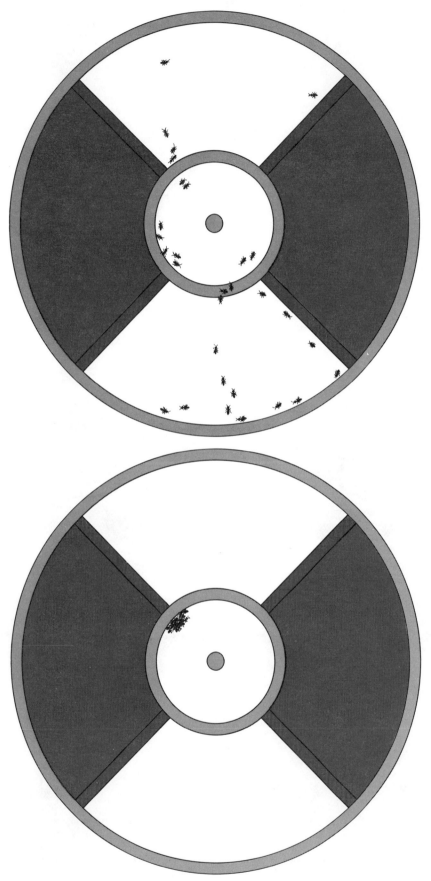

BEHAVIOR OF ANTS in the arena depended on the phase of the colony. When workers came from a colony in the early stage of a nomadic phase, they ran throughout the arena, spending equal time in light and dark areas (*top*). Statary workers usually went to one or both of the dark areas. In many tests, however, they remained in the cartridge (*bottom*).

ture adult ants to follow the trail substance of their colony.

To obtain the substance for the purposes of the experiment we washed ants from a colony in ether as a solvent. This procedure gave us an extremely potent solution of the substance that could be diluted to any desired strength by adding pure ether. An artificial trail was now deposited on a disk of chromatography paper that was rotated on a phonograph turntable as the solution flowed onto it from a microburette suspended above it. Within seconds after the circular trail had been deposited the solvent evaporated, leaving behind an invisible residue of the trail substance. The paper disk with the circular trail was removed from the turntable and placed on a template. The template had a black circle drawn on it that coincided exactly with the location of the invisible chemical trail on the disk of paper above it. Because the template circle was visible through the chromatography paper we could easily determine whether or not a test ant was indeed following the trail. Each ant was admitted to the circular trail through a tunnel, and its ability to detect and follow the trail was observed.

First we examined the ants' speed of running over the trail. As we had expected, callows taken from their colony soon after their emergence from the pupal cocoon were considerably slower than mature ants in following the trail. With each day of development in the nest the callows improved in speed on our test, and week-old callows were able to follow the chemical trail almost as rapidly as fully mature workers.

We then compared callows and mature adults on the basic ability to follow the trail, apart from the question of speed. Since the observations in the field and our speed tests had indicated that maturity was an important factor, we expected that the proportion of callows able to follow the trail without error would be considerably smaller than the proportion of adult workers able to do so. To our complete surprise it turned out that statistically the callows scored just about as high as the adults in the fundamental ability to follow the chemical trail of their colony. Evidently the callows were fully capable of recognizing the trail substance and their slowness must have been due to physical immaturity.

Thus we are still left with the question: Why do the young callows stay in the nest instead of going out on raids with other workers? There are several hypotheses to be considered. First, it has been noticed that much of the callows' time in the first few days is spent in intensive feeding on the nest's food supply. It is possible that this preoccupation with feeding could serve to keep them in the nest during the day. A second hypothesis, and one we plan to test, is that young callows are strongly attracted, probably more so than mature adults are, to stimuli of physical contact and chemical secretions that originate with other members of the colony. The intensity of both of these forms of stimulation is greatest inside the nest. A callow ant leaving the nest will experience a sudden reduction in the intensity of both classes of stimuli. Outside the nest the amount of tactual stimulation decreases as the adult ants fan out along the trail. In the outside air and on the narrow trails the colony's odors are also vastly diluted. As the mature adult workers depart on their massive daily raids, the tension of chemical and tactual attraction between the inside of the nest and the outside becomes less one-sided, but the concentration and pull of the inside might still be stronger. The direction of pull would be reversed only when the colony leaves the nest during an emigration to a new site. As the workers, the larval and pupal brood and the queen move out, the departing stimulation might attract the callows out of the nest. The decisive attractive force may be either the quantitative shift of the predominant mass of individuals or the departure of a source of stimulation that is particularly attractive to the callows, such as the queen, the developing brood, the total population of adult workers or the booty (the food supply).

We have recently made an interesting observation in the field that is consistent with the hypothesis that it is social stimuli that are responsible for keeping callow army ants in the nest. A colony of *Neivamyrmex nigrescens* was completing its statary phase in a bivouac located in the bank of a stream. One day there was a heavy rain followed by a flash flood that destroyed most of the colony; only a few hundred workers and fewer than 100 pupae were left. Under these conditions the callows ran along the entire route of the raiding column with the mature adult ants on the first night after their emergence from the pupal stage of development.

In many respects we have hardly scratched the surface in our attempts to understand how the social organiza-

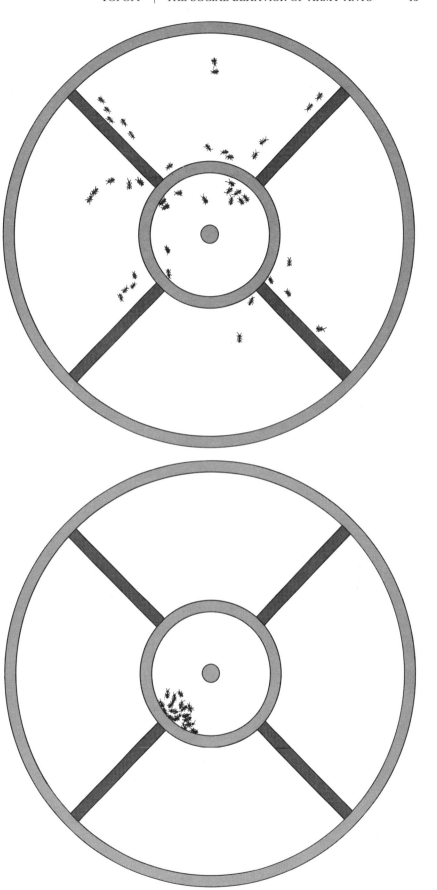

UNDER INFRARED ILLUMINATION the ants continued to behave as they had when they were exposed to visible light. Workers in the early nomadic phase ran rapidly through all the chambers of the arena (*top*) but statary workers often stayed in the cartridge (*bottom*).

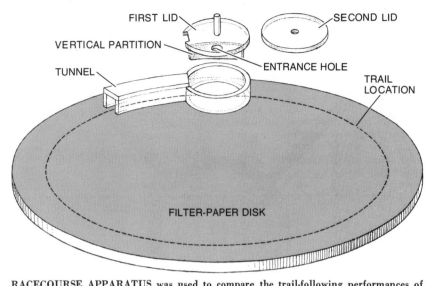

RACECOURSE APPARATUS was used to compare the trail-following performances of newly emerged "callow" workers with older workers. A circular scent trail was deposited on a disk of filter paper (*broken line*). Each ant was then placed in a chamber with a partition that prevented any contact with the trail. When the partition was removed, the ant could run along a short segment of the trail enclosed by a tunnel. On emerging the ant was scored for its ability to complete the circular course and for its mean running speed. By diluting the scent the experimenters were able to simulate both "strong" and "weak" trails.

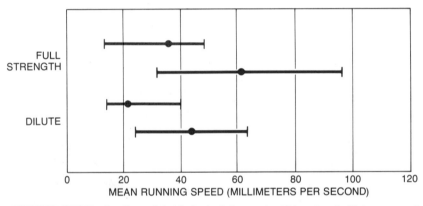

RUNNING SPEED of callow adults (*color*) of the species *Eciton burchelli* is compared with the speed of mature adults of the same species (*black*). Over a full-strength scent trail (*top bars*) the mature adult speed was better than 60 millimeters per second whereas the callow adult speed was little more than half that. Over a trail only a tenth as strong (*bottom bars*) the performance of both mature and callow ants was poorer but the gap was the same.

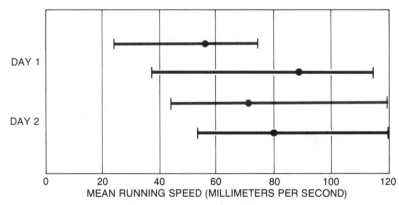

CALLOWS' PERFORMANCE improved as they matured. Graph compares the running speeds of callow (*color*) and mature adult (*black*) workers of the species *Eciton hamatum* on two consecutive days. On the first day the difference in mean speeds was some 30 millimeters per second (*top bars*). The next day the difference was less than 10 millimeters per second (*bottom bars*). Tests showed that the callows are virtually mature within seven days.

tion of army ants evolved, develops and is maintained by interactions among individuals, each of which has a limited capacity for behavioral adjustment. On the other hand, we have accumulated sufficient knowledge to enable us to compare processes that are important for the social organization of army ants with those that are important for the social organization of species representing other levels of evolutionary history. We know, for instance, that social organizations in all species of animals are maintained by physiological and behavioral interactions among the individuals that enable the group to function as an integrated unit. But each species of animal has an evolutionary and developmental history that is different from all others, and consequently each species has a unique morphology, physiology and behavior. This means that when the comparative animal behaviorist observes similar patterns of social behavior in two species of animals, he cannot automatically conclude that the mechanisms and processes underlying the behavior are the same for both.

For example, in both ant and human societies individuals exhibit different and specialized behavioral functions, giving rise to a division of labor within the group. But the role that any particular ant plays in its society is influenced directly by its biological organization, whereas human jobs are determined more by economic status, level of education, personal preferences and other cultural factors. Practically every behavior pattern that ants exhibit is based on their responses to a limited number of tactile and chemical stimuli; individuals in human societies interact by means of a much more complex form of communication based primarily on the use of symbolic language. Finally, although the behavior of every adult army ant is influenced by many social experiences it has during its development, the degree to which such developmental factors can modify adult behavior is certainly much smaller in ants than it is in humans. The goal of the comparative animal behaviorist is to study and clarify the bases for social behavior in species representing all levels of invertebrate and vertebrate evolutionary history. Only then shall we be able to judge how unique each species, including man, is. The study of the social behavior of army ants contributes to the attainment of this goal because it gives us a larger view of the diversity of social systems that are to be found within the animal kingdom.

Weaver Ants

by Berthold K. Hölldobler and Edward O. Wilson
December 1977

*These social insects use their own larvae as shuttles
to weave leaves into large nests in the rain forests of
Africa and Asia. Their behavior is coordinated by
complex chemical stimuli.*

Some insect species live in advanced social orders characterized by cooperation, caste specialization and individual altruism. Among the thousands of species of social insects a few deserve to be called classic, because certain remarkable features in their behavior have prompted unusually careful and thorough studies. The honeybees, the bumblebees, the driver ants, the army ants, the leafcutter ants, the slavemaker ants and the fungus-growing termites are all examples of classic social insects. The latest candidates for this select group are the weaver ants of the genus *Oecophylla* of Africa and tropical Asia. These ants devote a large part of their behavioral repertory to communication. The communication is further enriched by variations based on the weaver ants' caste system. As a result a weaver ant colony can perform feats far beyond the capabilities of single ants.

Weaver ants are extremely abundant, aggressive and territorial. They have achieved a position of exceptional ecological importance in the rain forests, cacao plantations and similar wooded environments they inhabit. For this reason the weaver ants have been the object of an increasing number of field studies. Over the past two years, however, we have succeeded in cultivating colonies of the African species *Oecophylla longinoda* in the laboratory. We induced the ants to live in potted trees and glass tubes. Under these conditions it has been possible for the first time to study the full range of the social life of the weaver ant.

These slender yellow insects are exquisitely adapted for life in the leafy canopies of tropical forests. Their main social unit is the colony, which consists of as many as 500,000 female workers, the progeny of a single enormous queen. The caste system within each colony consists of three forms of adult female: the heavy-bodied queen, a large population of "major" workers and a smaller population of "minor" workers. The weaver ant males, like those of other

ant species, participate relatively little in the social life of the colony. They leave soon after maturing to participate in wedding flights with the virgin queens, after which they die without returning to the nest.

The major workers are fairly large, averaging six millimeters (about a quarter of an inch) in length. They are the general laborers, responsible for most of the foraging and nest construction. The more aggressive of the two worker castes, they rush from the nest at the slightest disturbance to bite an intruder and release formic acid from their poison gland. Major workers also form a dense retinue around the queen. They grasp her with their powerful legs so tightly that at times she is held in midair in the center of the nest cavity. About once a minute one of the major workers regurgitates a liquid meal into the mouth of the queen. At somewhat less frequent intervals a member of the queen's retinue lays a special trophic egg—a flaccid object without the ability to survive—that is immediately fed to the queen. This virtually continuous flow of nutrients enables the queen to manufacture hundreds of eggs a day. As the eggs are extruded from the queen's oviduct major workers carry them to special brood piles. There the smaller minor workers care for the eggs and feed and wash the tiny larvae that hatch from them. When the larvae are close to their maximum size, the major workers and minor workers share about equally in their care.

Weaver ants are named for their method of nest construction. The nests are made of leaves folded or fastened together to form tight, tentlike compartments. The leaves are held in place by seams of silk spun by the larvae, which the major workers employ like shuttles for weaving the nests. This nest building is one of the most remarkable instances of social cooperation among lower animals.

Once the weaver ants have chosen a

tree branch suitable for a nest they spread out on the leaves of the branch and begin to pull on the tips and edges. When an ant succeeds in turning up a segment of a leaf, nearby workers are attracted to that part of the leaf, and soon there is a small group of ants pulling in unison. When a leaf is broader than the length of an ant's body, or when two leaves must be pulled together across a wide space, the workers form living bridges between the points to be joined. Then some of the ants in the chain climb onto the backs of their neighbors and pull backward, thus shortening the chain and bringing the leaf edges together. When the leaves have been maneuvered into shape, some of the ants remain on them, employing their legs and mandibles to hold the leaves in place. Other ants go back to already established nests and return to the new site carrying partly grown larvae. The workers wave the larvae back and forth across the leaf seams. This causes the larvae to release threads of silk from gland openings located just below their mouth. Thousands of these threads woven into sheets are strong enough to hold the leaves in place. Sheets of silk are also spun to make circular entrances and outer galleries leading to the interior of the new nest.

A single weaver ant colony can occupy a substantial volume in the canopy of a forest. The colony can fill an entire tree or even several adjacent trees without breaking the lines of communication that are vital to social insects. From the leaves of the trees the weaver ants construct hundreds of nests to serve as retreats, nurseries and outposts. During the day foragers patrol every square centimeter of leaf and bark within their territory. They rout enemies, capture insect prey and gather the sweet "honeydew" excrement of the swarms of scale insects and other sap-feeding homopterous insects that the ants guard as though they were dairy cattle.

The species of *Oecophylla* are not the only insects that weave. A few other tropical species use larval silk to construct nests in trees and shrubs. The weaver ants are distinguished, however, by their close control of their environment. Indeed, man has employed weaver ants to control the arboreal environment for him.

Records from the Canton area of China show that weaver ant nests were gathered, sold and placed in selected citrus trees to combat insect pests in about A.D. 300. The same technique was noted in the 12th century and was still practiced in southern China well into the 20th century. The weaver ant used for this purpose is the Asian species *Oecophylla smaragdina*. This utilization of weaver ants is the oldest-known instance of the biological control of insects in the history of agriculture. Recently Dennis Leston, formerly of the University of Ghana, and other entomologists have recommended employing the African species of weaver ant to control pests of tree crops such as cacao. Studies in Ghana have shown that the presence of weaver ants reduces the incidence of two of the most serious diseases of cacao, one caused by a virus and the other by a fungus. In both cases the pathogen is transmitted by mirid leaf bugs. The weaver ants evidently combat the diseases by attacking the bugs. The *Oecophylla* workers are also particularly effective in hunting insects that feed on the tissue and sap of trees.

The weaver ants' exceptional control of their environment has been achieved through the evolution of advanced forms of social behavior. The communication system we have observed in our studies of the African species of weaver ant is one of the most complex and advanced systems known among the social insects. The great strength of the weaver ants lies in their ability, demonstrated in their nest building, to cooperate in group activities. They employ five different recruitment systems, consisting of distinct combinations of chemical and tactile signals, to initiate other group endeavors. These recruitment systems are employed in the main occupations of the weaver ants outside the nest: penetrating new territory, defending it and extracting food from it.

Weaver ants have an impressive sense of place that helps them to secure new territory. Their large eyes give them vision that is unusually acute for ants. Moreover, they are able to remember the appearance of many details of the nest area. If an object is simply shifted from one side of a nest tree to the other, workers come out of the nest to explore the object as if it were fresh terrain. In fact, when a conspicuous object such as a potted plant or a box is moved close to a weaver ant nest, the alert workers crawl out over the branches and leaves of their own tree in an effort to reach it. If the ants fail to get onto the new surface by reaching with their legs, they begin to climb on top of one another, constructing pyramids or chains with their bodies until the chasm is bridged. Then workers rush onto the new territory and begin to explore it.

The first explorers return to the nest to recruit other workers to help in securing the new territory. They mark the route from the territory to the nest with odor trails, that is, trails of a pheromone, or message-bearing chemical, that will guide their nestmates to the new area. The odor trails are created in an unusual way. The ants extrude a glandular segment of the hindgut through the anus. This organ, which we call the rectal gland and which is known only in weaver ants, is used in four of the recruitment systems. When the gland is extruded, it rests on a tiny sled consisting of two bristles that project from the tip of the ant's abdomen. As an ant runs back to the nest a secretion from the rectal gland is brushed on the ground, thereby creating the odor trail. When an ant laying one of these trails encounters nestmates, it jerks its body in their direction while touching them on the head with its antennae. The greeting stimulates the nestmates, with the result that they follow the trail to the new territory and begin to explore it. We call this process the system of recruitment to new terrain.

LIVING CHAIN of weaver ants folds over the tip of a leaf during the construction of the arboreal nest, a remarkable example of cooperative behavior. The workers shorten the chains by climbing up on the backs of nestmates while hauling edge of the leaf behind them.

CASTE SYSTEM of the African weaver ant consists of three forms of adult female: a single queen, a population of "major" (large) workers that forage for food and perform a variety of other tasks, and a lesser number of "minor" (small) workers that care for the eggs and younger larvae. The queen (*center*) is occupied solely with receiving food from the major workers, chiefly by regurgitation, and laying eggs. In the foreground one major worker regurgitates to a larva; a second lays a "trophic" egg that lacks survival ability and will be fed to the queen. Minor workers are clustered around a brood pile. Scene was painted by Turid Hölldobler. Ants are enlarged about six times.

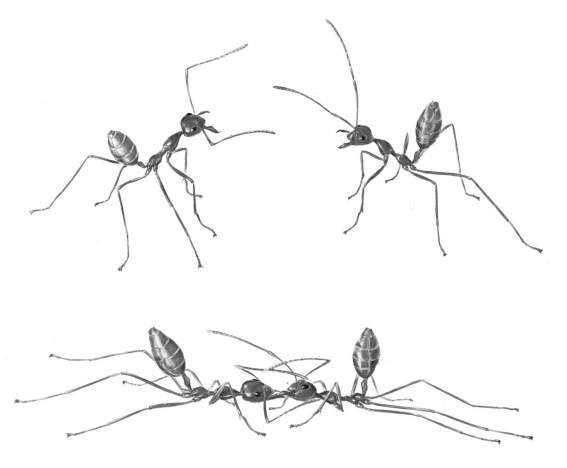

TERRITORIAL COMBAT of two weaver ants is initiated by a dance-like maneuver in which the combatants raise themselves by extend-ing their legs and circle each other with stiff, jerky movements (*top*). Then one ant attempts to seize the other with its mandibles (*bottom*).

SILK TO BIND NESTS is spun by partially grown larvae, which the adult workers hold in their mandibles and move back and forth over the leaf seam like shuttles. The threads of silk, released from gland openings just below the larva's mouth, are woven into sheets.

COMPLETED NEST was constructed by a colony of the African weaver ant (*Oecophylla longinoda*) in a potted grapefruit tree grown in authors' laboratory at Harvard University. In the wild a colony of some half-million workers (the progeny of a single queen) may construct hundreds of nests from the leaves of one or more trees. The ants patrol this domain by day and withdraw into the nests at night.

After an extension of a weaver ant colony's territory has been secured, the workers sweep back and forth across it in search of food. When a worker encounters a sugary secretion (usually from scale insects), it returns to the nest to recruit nestmates. Once again the rectal gland is employed to lay an odor trail. In this recruitment system the worker stimulates its nestmates by stroking them with its antennae while offering them regurgitated food from the find. Emigration to nests built in the new territory is effected by still another recruitment system, one that includes a rectal-gland odor trail, stroking with antennae and the physical transport of nestmates.

Two more recruitment systems are employed in defending the colony's territory. Weaver ants are particularly aggressive toward members of other weaver ant colonies. In fact, Leston has found that the territories of different colonies are separated by "no-ant's-lands," that is, narrow zones into which few ants venture. A contact between colonies of weaver ants usually results in immediate, spectacular warfare, numerous casualties and eventually the retreat of one of the colonies from part or all of its territory. In the natural habitat of the African species such battles can last for days as the massed opponents struggle along slowly shifting lines of defense.

Foraging workers that encounter enemy weaver ants react with a series of swift, precise movements. Individual combat is initiated with a dancelike maneuver in which the combatants raise themselves on extended legs and circle each other with stiff, jerky movements. Then they thrust and snap at each other with their mandibles. A defeated ant is pinned spread-eagled to the ground. Its legs and antennae are then clipped off and scattered, and its abdomen is often sliced open as well. Throughout the melee ants crumple dead and dying. Some of the workers rush back to the nest laying rectal-gland odor trails. When the trail layers encounter nestmates, they jerk their bodies in what appears to be a ritualized version of the preliminary combat dance. The nestmates respond, however, not by fighting but by running out along the trail to the battle site.

At the same time the fighting workers employ a shorter-range recruitment system to organize group attacks. When a forager encounters an enemy ant but fails to engage it in combat, the forager often runs short looping patterns while dragging its abdomen over the ground. In this recruitment system the ant rotates the terminal segment of its abdomen to expose its sternal gland. This newly discovered organ is also known only in weaver ants. The ants are attract-

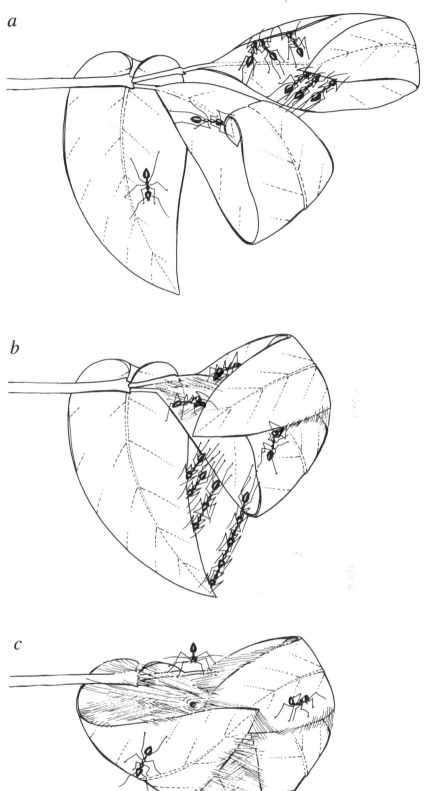

COOPERATIVE NEST BUILDING in the weaver ant is illustrated in this sequence. At first the workers labor independently in their attempts to pull down or roll up leaves. When success is achieved by one or more of them at any part of the leaf, other workers in the vicinity abandon their own efforts and join in (*a*). When the leaves have finally been shaped into tentlike configurations, some of the ants continue to hold them in place with their legs and mandibles while others carry partially grown larvae (*color*) from preexisting nests and bind leaves together with sticky larval silk (*b*). Sheets of silk are then added to create circular entrances and galleries (*c*).

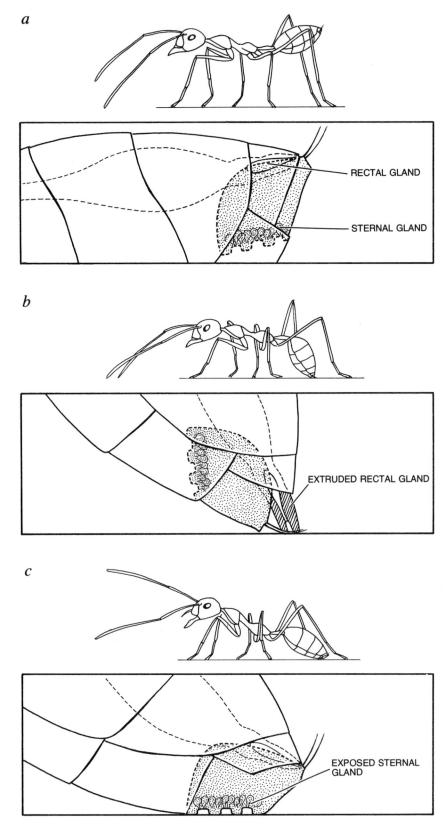

ODOR TRAILS are laid down by weaver ants to coordinate social activity. Normally the foraging worker ant walks with its abdomen elevated (*a*). When the ant encounters a new terrain or food source, it lowers its abdomen, extrudes the rectal gland through the anus and deposits a pheromone, or message-bearing substance, in a line along the ground (*b*). While the rectal gland is extruded it rests on a tiny sled consisting of two bristles. Nestmates that have been tactilely stimulated by the trail-laying ant will follow the trail to a new terrain. Short-range recruitment of major workers to fight invading ants and other intruders is achieved by exposing the sternal gland of the abdomen and dragging it over the ground to deposit a short, looping odor trail (*c*).

ed by the scent of the sternal-gland secretion from as far away as 10 centimeters. As a result small clusters of ants gather at the places where the enemy was first encountered.

J. W. S. Bradshaw and his colleagues at the University of Southampton have shown that this effect is enhanced by the release of various alarm substances from glands located at the base of the weaver ant's mandibles. Some of these substances attract nestmates to the scene. Others increase the ants' level of excitement and aggressiveness. We have observed repeatedly that these groups of recruited ants are far more effective in combat than individual ants. For example, workers of the large black tree ant *Polyrhachis militaris* can easily knock aside single weaver ants. When three or more weaver ants form a tight group, however, they can seize a tree ant in concert and pin it to the ground. Other workers quickly converge on the spot and assist in the kill.

The weaver ant is a sophisticated judge of odor cues. We noticed that after foragers occupied a new territory in the laboratory they began to deposit large droplets of fecal matter over the surface of the territory. This behavior differs from that displayed by most kinds of ants, which concentrate their excrement in refuse areas and other restricted locations. When the weaver ants patrol their territory, they inspect the fecal droplets. If a weaver ant encounters a droplet left by a member of another colony, it reacts momentarily with aversion, assumes a hostile posture and then inspects the droplet more closely. We have been able to induce the same set of reactions with fluid taken from the hindgut of alien ants.

The fecal substances give weaver ants an advantage when they are the defenders in territorial combat. We arranged a series of eight "wars" between colonies in areas previously marked out with fecal droplets by one or another of the colonies. In each case the members of the colony that had deposited the droplets were less hesitant to forage over the terrain and quicker to recruit nestmates when they encountered alien ants. As a result they gained the initial advantage and secured more ground during the initial fighting.

It appears that the distinguishing characteristics of the weaver ants evolved a long time ago. The African and Asian species are survivors of one of the most distinctive and ancient lineages of ants. Many insect fossils from over the past 100 million years are preserved in amber, which is fossilized resin. Fossils of two extinct species of weaver ant, *Oecophylla brevinodis* and *Oecophylla brischkei*, are found in amber that was deposited in the area of the Baltic Sea some 30 million years ago in the Oligocene epoch. During that period northern Europe contained both tropical and Temperate Zone forests. The fossil record shows that many of the insects of those forests resembled the insects of similar environments in the Europe and tropical Asia of today. In particular the extinct species of *Oecophylla* are related more closely to the modern Asian species than to any ants that are now found in Europe.

Some 15 years ago a first glimpse was obtained of the social organization of the extinct *Oecophylla*. In 1963 Mary Leakey was searching for fossils on Mfwangano Island in Lake Victoria in Kenya. She uncovered an assemblage of ant fossils: 366 tiny crystalline ants clustered in a single spot. The ants apparently had been living in a leaf nest that had fallen into a freshwater pool, where the nest and its inhabitants were quickly covered with sediment. Under those conditions an unusual amount of surface detail was preserved. One of us (Wilson) identified the assemblage as a portion of a colony of extinct weaver ants. It is the first and so far the only

insect society that has been found in the fossil state. The assemblage includes clusters of larvae and pupae. Some of the fossil ants are still attached to fragments of leaves. Since there was a small population rather than the usual single fossil specimens, it was possible to make a statistical study of the caste system of the fossil species. The anatomical characteristics of these ants and the relative abundances of the two worker castes turned out to be quite similar to those that, among modern ants, are unique to the *Oecophylla* species.

The most distinctive element of the caste system of modern weaver ants is the minor worker: the caste of smaller and less numerous ants specializing in the care of eggs and small larvae. In most ant species that have more than one caste it is the major workers that are anatomically deviant and less numerous. The fossil population discovered by Mary Leakey was in a lower Miocene deposit and is therefore at least 15 million years old. The unusual anatomy and size distribution shared by the Miocene and modern weaver

ants suggest that the peculiar division of labor among the living species of *Oecophylla* is of great age.

As the evidence from the fossil weaver ants implies, advanced social organization confers an evolutionary stability on the insect societies. This stability has advantages and disadvantages. On the one hand our studies suggest that individual workers perform no more than 50 distinct behavioral acts, most of which are for the purpose of communication. The result of this almost exclusively social orientation is that the weaver ant colony is an extremely successful working unit. On the other hand, the effectiveness of the colony is obtained by the rigid programming of the relatively simple components of individual behavior, which secures a complex but lockstep pattern of cooperation during group activities. It appears that the colony can flourish only at the expense of any semblance of independent action on the part of the individual. In other words, the weaver ants, like the honeybees, seem to have reached the extreme of one spoke of adaptive radiation among the social insects.

SYSTEM	FUNCTION	CHEMICAL SIGNALS	TACTILE SIGNALS	PATTERN OF MOVEMENT
Recruitment to food	Recruitment of major workers to immobile food source, particularly sugary materials	Odor trail from rectal gland and regurgitation of liquid from crop	Touching with antennae, head-waving and mandible-opening associated with the offering of food	Looping trails laid around food source, with main trail leading directly to nest
Recruitment to new terrain	Recruitment of major workers to new terrain	Odor trail from rectal gland	Touching with antennae and occasionally jerking body back and forth	Broad, looping trails laid around new terrain with deposition of hindgut material containing territorial pheromone; main trail leads directly to nest
Recruitment during emigration	Emigration of members of colony to a new nest site	Odor trail from rectal gland	By touching with antennae, one ant indicates its readiness to carry another to new nest site	Main trail leads directly to nest site without additional looping trails. Workers carry first mostly larvae and pupae, then other workers to nest site
Short-range recruitment to enemies	Short-range recruitment of nestmates for assembly and more rapid capture of invaders and prey	Short, looping odor trails from sternal gland; exposure of gland surface when abdomen is lifted in air	None	Short, looping trails limited to the vicinity of contact with enemy
Long-range recruitment to enemies	Long-range recruitment of major workers to fight invaders. Particularly intense during territorial wars with members of same species	Odor trail from rectal gland	Touching with antennae. During periods of greater excitement, body is jerked back and forth	Main trail leads directly to nest

RECRUITMENT SYSTEMS of the weaver ant are summarized in this table. The two pheromones secreted by the rectal gland and the sternal gland, when combined with tactile signals and the spatial configuration of the odor trail, can communicate five different messages.

6

The Schooling of Fishes

by Evelyn Shaw
June 1962

*What influences make a fish join others of the same
species to form a school? The question is studied partly
by observing the developing behavior of young
schooling fishes in a special laboratory aquarium.*

For sea gulls, fishermen and other predators the propensity of certain species of fish to assemble in large schools is a great convenience. A school of fish is something more, however, than a crowd of fish; it is a social organization to which the fish are bound by rigorously stereotyped behavior and even by anatomical specialization. Schooling fishes do not merely live in close proximity to their kind, as many other fishes do; they maintain, during most of their activities, a remarkably constant geometric orientation to their fellows, heading in the same direction, their bodies parallel and with virtually equal spacing from fish to fish. Swimming together, approaching, turning and fleeing together, all doing the same thing at the same time, they create the illusion of a huge single animal moving in a sinuous path through the water.

This peculiar social organization has no leaders. The fish traveling at the leading edge of a school frequently trade places with those behind. When the school turns abruptly to the right or left, the fish on that flank become the "leaders," and what was the leading edge becomes a flank. Except in the execution of such a turn and during feeding—when the school formation may break up completely—the fish swim parallel to one another. The distances between fish may vary as individuals swim along at different and changing speeds, particularly in a slower moving, loose school. When a school is startled, for example, by a predator or an observer, it closes ranks immediately and the fish-to-fish spacing becomes equal and fixed as the entire school takes flight.

Even in schools of as many as a million fish, all members are of a similar size. Speed increases with size and the fish of a species therefore tend to sort themselves out by size and by genera-

tion in the sea. Schools can take many shapes and usually have a third dimension, being a few fish or many fish deep. From above they may appear rectangular or elliptical or amorphous and changeable. Some species form schools of characteristic shape. The Atlantic menhaden, for example, can be easily recognized from the air because their schools move through the water like a giant amoeboid shadow, often changing course but never breaking apart.

The speed and synchronization of response, the parallel orientation and the constancy of spacing among members of a school inevitably suggest that their behavior is integrated by some central control system that makes each "think" of changing course at exactly the same moment. Of course, there is no such central control system. Nor is it possible to explain the simultaneity of the members' actions as response to external stimuli from the environment. From time to time the fish do respond, as other animals do, to such stimuli as food and change of light intensity. Environmental conditions, however, do not explain the high degree of synchronized parallel movement that the members of a school display moment after moment, day after day. In fact, the great stability of schools, persisting through the most varied environmental conditions, suggests that the school organization must be dominated by internal factors.

Schooling is easily enough explained as an instinct. The term implies a causal factor—saying, in effect, that fishes school because they have an instinct to school. This tautology does not explain much, even when it is amplified by the more sophisticated statement that the behavior is inborn, unlearned and characteristic of the species. Many animals exhibit clear-cut, species-specific pat-

terns of behavior, and it is useful to seek these out and compare them as they appear in related species. Such inquiry leaves equally interesting questions unanswered. In the present instance it does not explain what brings about the concerted action of the fish in a school. This requires, above all, study of the behavior as it unfolds in the developing organism. With growth and particularly with the maturation of the sensory system, the relation between the organism and its environment changes. The life history of the individual, however typical of its species, has a profound role in the molding of the behavior of the mature animal and holds the principal clues to the mechanism that governs its interaction with its social and physical environment. So far this approach to the schooling of fishes has only made the mystery more intriguing.

With progress on the question of how fishes school, one can also hope for some light on why fishes school. No other line of study has disclosed what function this highly organized social behavior serves in the perpetuation of the species that have adopted it.

In my own work at the Marine Biological Laboratory at Woods Hole, Mass., at the Woods Hole Oceanographic Institution, at the Bermuda Biological Station and at the Lerner Marine Laboratory on Bimini in the Bahamas, I have attempted to overcome the difficulty of study in the field by bringing fishes into the laboratory for observation and experiment. Life begins for most species of schooling fish in the plankton, where the eggs drift untended and abandoned by the school that laid and fertilized them in its passage. The eggs develop into embryos and the embryos into larvae, or "fry," which are capable of some feeble swimming movement. They grow, they ma-

SCHOOL OF HERRING was photographed by Ron Church near San Diego, Calif. The majority of herring caught in the Pacific Ocean are used to make fish oil and fish meal. This school, originally headed straight for the camera, has begun to turn to its right.

SCHOOL OF MULLET, which are common in the waters off Florida, was photographed there by Jerry Greenberg. A member of the order Mugiliformes, the mullet is an oceanic fish, and its distribution is primarily on both sides of the temperate South Atlantic.

ture and at some point during their early lives come together and form schools. One would like to be able to observe them during this epochal period. The only way to find the fry in the open oceans is to gather them in a plankton net, and the net necessarily disrupts the normal pattern of their behavior. My field studies have therefore been restricted to species that can be found as fry near the shore. But the fry are so tiny that crucial stages in the unfolding of their behavior in their natural habitat must go unseen.

In the waters around Cape Cod I have worked in particular with two species of *Menidia*, known commonly as whitebait, spearing or silversides. During late spring and early summer they spawn heavy eggs that adhere by sticky threads to rocks and to the stems of marine grasses and algae. On hatching, when they are no more than five millimeters (about a quarter of an inch) in length, they become part of the plankton. In spite of patient search I have never observed fry this small in open waters. When they grow to seven millimeters or longer, they become easier to find in the plankton. I have seen fry seven to 10 millimeters long randomly aggregated in groups but not yet schooling or showing any sign of parallel orientation to one another. As the season progresses and as they grow from 11 to 12 millimeters in length, they can be observed forming schools for the first time, lining up in parallel, with 30 to 50 fry to the school. During the summer of 1960 my associates and I observed an estimated 10,000 of these tiny fishes in the plankton of the shallow waters near Woods Hole and collected many of them.

From these observations one could deduce that schooling begins when the fry reach a certain length. We could not tell, however, whether schooling develops gradually or happens suddenly. We accordingly proceeded to rear some 1,000 *Menidia* from the egg in the laboratory. For the study of these fry we set up a doughnut-shaped tank with a channel three inches wide, having observed that schools tend to break up when they approach the corners of a rectangular tank. We took care also to observe them in constant light and through a one-way mirror. We were reassured to find that under these condi-

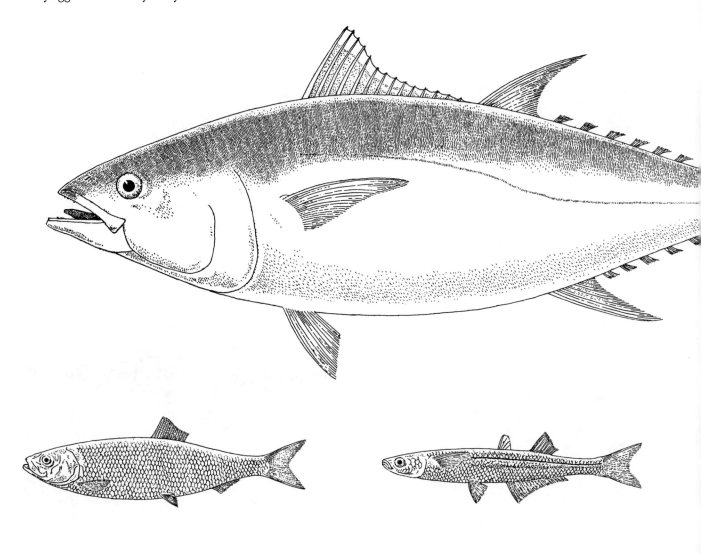

REPRESENTATIVE SCHOOLING FISH shown on these two pages are a tuna (*Thunnus thynnus*), at top left; a herring (*Clupea harengus*), at bottom left; a silverside (*Menidia menidia*), second from bottom left; a mackerel (*Scomber scombrus*), at top right; and a

tions schooling appeared in laboratory-reared fry when they grew to the same size as the smallest schooling fry observed in the sea.

The close-up and constant surveillance in the laboratory showed that schooling unfolds gradually in characteristic patterns of fish-to-fish approach and orientation. Newly hatched fry, five to seven millimeters in length, would approach the head, the tail or the side of other fry to within five millimeters and then dart away. At eight to nine millimeters in length, a fry would approach the tail of another fry and, when the two fry were one to three centimeters apart, they would swim on a parallel course for a second or two. If either fry approached the other head on at an angle, however, each would dart off rapidly in the oppo-

site direction. At about nine millimeters in length the head-to-tail approach became predominant, and the fry would now swim on parallel courses for five or 10 seconds. When they reached a length of 10 to 10.5 millimeters, one fry would approach the tail of another and both fry would briefly vibrate their entire bodies. This curious behavior would terminate with the two fry swimming off in tandem, or in parallel, for 30 to 60 seconds, occasionally joined by three or four other fry in the formation of a recognizable little school. The number that would engage in this behavior increased to 10 or so when the fry reached a length of 11 to 12 millimeters. With the distances from fish to fish ranging from 10 to 35 millimeters, the school was a ragged one. By the time the fry

had grown to 14 millimeters the fish-to-fish spacing became less variable, ranging from 10 to 15 millimeters, and there was less shifting about in the school.

Schooling behavior can therefore be described as developing initially from the interaction of two tiny fry. As they grow older and larger, the head-on approach gives way to the head-to-tail approach; the two fry tend to swim forward in parallel instead of fleeing from one another, and they are joined by increasing numbers of individuals in the formation of the incipient school.

At this point some speculation is in order, particularly if it suggests specific hypotheses for exploration by observation and experiment. During the head-on approach, one may suppose, each fry sees a changing visual pattern:

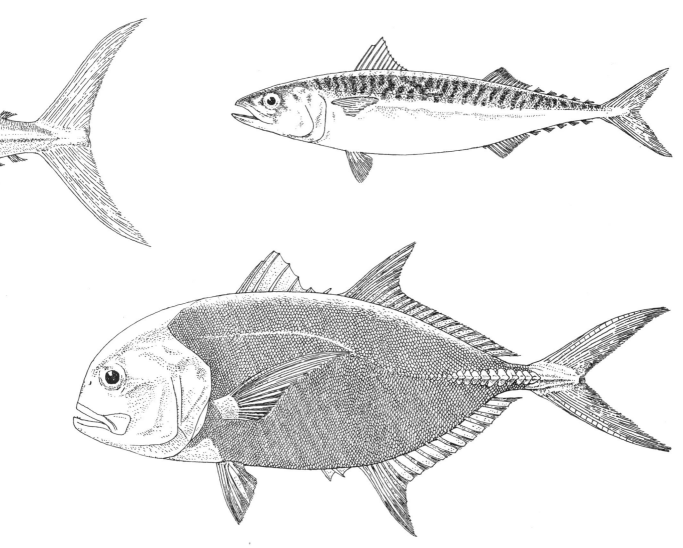

jack (*Caranx hippos*). The fish are not drawn to scale. Tuna have been known to reach a length of 14 feet. The mackerel averages 14 to 18 inches, and the jack about two feet. The silverside grows to six inches; the herring may reach a length of 12 inches.

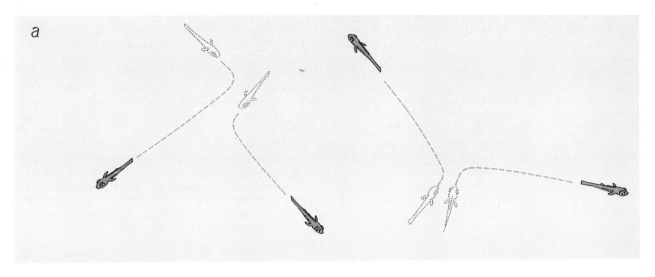

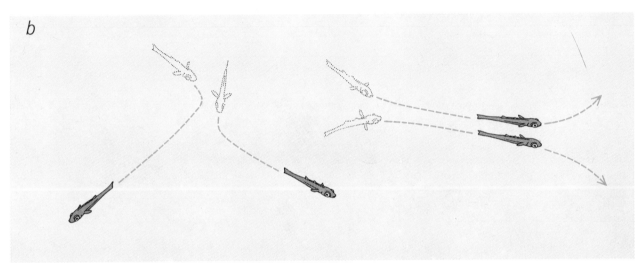

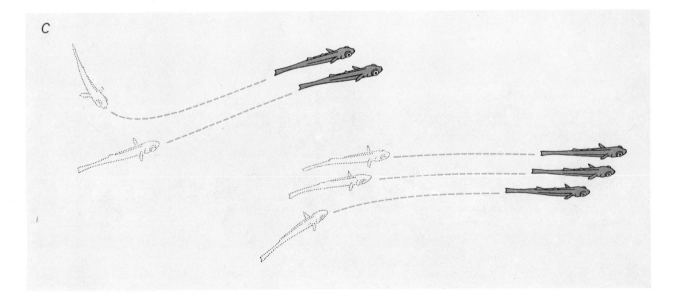

SCHOOLING ACTIVITY OF JUVENILE FISHES, or fry, develops as they grow. When newly hatched fry five to seven millimeters in length (*top*) approach the head, tail or side of other fry to within five millimeters, they dart away. At eight to nine millimeters (*middle*) two fry school momentarily if one has approached the tail of the other, but a side approach or one to the head still makes them dart away. As the fry grow from a length of about nine millimeters to 10.5 (*bottom*), the head-to-tail approach becomes predominant and two fry will school for five to 10 seconds; they later begin to school for short periods in threes and fours.

an oval mass (the head) and bright black spots (the eyes) coming closer and closer. The stimulus becomes too intense and each fry veers off in flight. The tail-on approach, in contrast, presents a quite different, although changing, pattern. This time it is a small silvery stripe and a transparent tail, swishing rhythmically and steadily moving away. The approaching fry follows. The leading fry may see, out of the rear edge of its eye, only a vague image of the follower. In each case the visual stimulus is moderate to weak in intensity, and the two fry swim forward together.

T. C. Schneirla of the American Museum of Natural History has postulated that, in general, mild stimuli attract and strong stimuli repel, and that most animals tend to approach the source of a mild stimulus and withdraw from the source of a strong one, even if they have had no prior experience with these conditions. Our fry had had considerable time to accumulate experiences of mutual encounter. We could not be certain, however, about the nature and impact of such experiences. A natural question therefore arose: Is such experience essential to the nature of schooling behavior? Or, to let the question suggest an experiment: Will fishes show schooling if they are taken away from their species-mates and raised in isolation? One must be cautious, however, in interpreting the results of such an experiment. On finding that a given behavioral trait appears in an animal that has been reared in isolation, some students of animal behavior are ready to conclude that the trait must be innate or instinctive and to close the book on further investigation at that point. Perhaps the pitfall lies in the word "isolation." No animal can grow up in a total vacuum of experience. In the case of the fry we proceeded to rear away from their species-mates, it was clear that each one had experience of itself (although we coated the bowls with paraffin so that the fry could not see their own reflection), of the water in its bowl, of the *Artemeia* shrimp on which it dined and of such stimuli as reached it from the world outside its bowl.

The mortality among the fry we isolated in this fashion proved to be extremely high. Only four out of 400 survived to schooling size in the first season and only nine out of 87 in the second. Apparently the fry need one another in the earliest larval stage, but we do not yet know why. The one noticeable difference between those reared in the community of their fellows and those reared alone seems to show up in the initiation of their feeding behavior. The fry in our laboratory communities began to feed two or three days after hatching, while they still carried a large yolky sac on their abdomen, whereas their siblings in isolation evidently starved to death. When we placed fry in isolation a week after hatching and after they had begun feeding, we secured a somewhat higher survival rate and, it turned out, a different and still enigmatic result when it came to observing the emergence of their schooling behavior.

As soon as the first four fry reared in isolation reached schooling size, we placed them in the company of schooling fish in community tanks. At first they showed disorientation; they bumped into their species-mates and occasionally swam away from the school. At the end of four hours, however, these fry could not be distinguished in behavior from the others. What this experiment showed is that fishes reared in isolation will soon join a school. It did not answer the question of whether or not schooling behavior would appear in fishes so reared.

With a more adequate supply of fry reared in isolation and in semi-isolation during the summer of 1960, we found that they would indeed form schools. The fry that had never had any contact with species-mates schooled within 10 minutes after being placed together in the test chamber. Those that had spent the first week after hatching in the company of species-mates also formed schools, but it took some of them at least 150 minutes to do so. What is more, we found that the shorter the time they had spent in isolation, the longer it took them to form a school. This suggests that their early experience with species-mates—at the period when the fry are still approaching one another at odd angles and darting away—may have set up some inhibitory process.

Although these experiments indicate that isolation in infancy does not keep these fishes from forming schools, the role of experience deserves further study. In this connection it should be added that schooling behavior was established

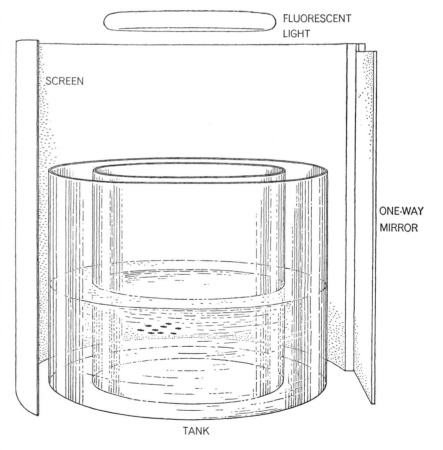

FISH TANK used by the author to study the development of schooling behavior in *Menidia* fry is doughnut-shaped and has a three-inch-wide channel in which the fry can swim continuously without reversing direction. The tank is completely encircled by a screen (*here cut away*). The fry are observed either from above or through the one-way mirror.

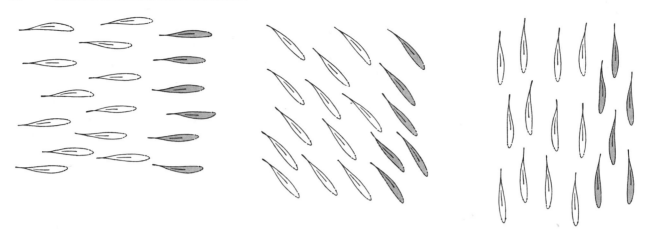

TURNING OF SCHOOL makes the relative position of the leading fish with respect to the school change. These fish (gray), which are originally at the leading edge (left), gradually shift around to the flank as the school turns (middle and right).

in our control communities when the fry were still a good deal smaller than the size at which we exposed our few precious isolates and semi-isolates to one another's company.

Another set of experiments with our laboratory fry produced evidence that the visual attraction of one fry for another develops in parallel with the emergence of schooling behavior. Very young fry showed no response at all to another fry swimming on the other side of a glass barrier. As the fry approached schooling age and size, however, they responded more and more actively to the visual image of the other fry. Finally they began to orient themselves in parallel to the fry on the other side of the barrier and were even observed to vibrate their bodies as they did so.

In a similar experiment with adult schooling fishes the visual attraction of one for another becomes readily apparent. Placed on each side of a glass partition, they swim toward each other immediately. In fact, fishes that cannot see cannot school. A fish blinded in one eye approaches and lines up with another fish on the side of the intact eye; a pair of fish blinded in different eyes swim at random when their sightless eyes are turned toward each other and school normally when they approach on the side with sight.

Just what visual cues are decisive in the mutual attraction of schooling fishes remains to be determined. Various experiments have shown that movement is important and that movement outweighs color and species, especially in attracting the initial approach. Albert E. Parr, now at the American Museum of Natural History, proposed in 1927 that fish-to-fish distances in schools might be explained by a balance of visual attraction and repulsion. According to Parr, the fish are repelled when they come too close together and attracted when they swim too far apart; the typical spacing in the school would thus represent the equilibrium of these two forces.

In a study of the schooling species around Cape Cod, Edward R. Baylor of the Woods Hole Oceanographic Institution and I found that many of these fishes are farsighted and that their retinas are therefore presented with a somewhat fuzzy image. The distribution of rod and cone cells in their retinas indicates, on the other hand, that their eyes may be well adapted for enhanced perception of contrast and so of motion against the hazy underwater background. This kind of vision would be highly adaptive in schooling behavior. Baylor and I also tried to modify the fish-to-fish schooling distance in pairs of fish by placing contact lenses over their eyes, but we observed no conclusive effects.

Although it appears that the visual apparatus is dominant in schooling behavior, there is also evidence that it does not serve as the exclusive channel of mutual attraction among fish. M. H. A. Keenleyside of the Fisheries Research Board of Canada has observed, for example, that Pristella, a species that sometimes schools, would respond to fish on the other side of a glass barrier by swimming back and forth along the barrier but would gradually lose interest, wandering away from the barrier more and more frequently and finally not returning at all. Sensory cues other than visual ones are most likely involved in establishing the parallel orientation and the fish-to-fish distances that give the school its ordered structure. It is difficult to determine which cues, because the experimenter cannot control for vision—a fish deprived of sight cannot make the initial approach so essential to the rest of the process.

Hearing, taste and smell have all been implicated, although inconclusively. James M. Moulton of Bowdoin College found that different schooling species produce different sounds, mostly of hydrodynamic origin, as the fish stream and veer through the water. Such sounds, in Moulton's opinion, may help to maintain the total school. There is no evidence, however, that sound helps to keep an individual fish oriented in position in the school. There is even less to be said for taste and smell, particularly in the case of oceanic fishes. Such odors as the fishes might produce would be diluted in the sea; although they might act on individuals at the trailing edge of a school, they could play little role in the behavior of those in the vanguard.

The one sensory system that would seem to be designed to play a role in the orientation of fish to fish is that associated with the lateral line—the nerve and its associated branches that are distributed over the head and run from head to tail along each side. It is thought that this organ is responsive to vibrations and water movements. Willem A. van Bergeijk and G. G. Harris at the Bell Telephone Laboratories have reported evidence indicating that the lateral line is sensitive particularly to "near field" motion of the water produced by propagated sound waves. Parallel orientation may well be facilitated by information about the movements of nearby fish picked up by the lateral line. The approach of one fish to another induced by visual attraction might also be checked by the increasing force of lateral-line perceptions of the movements of the same companion at closer range.

That schooling is a successful way of life can be judged from the fact that so many fishes have adopted it. Some 2,000 marine species school, and there is a major group—the Cypriniformes, consisting mainly of fresh-water fishes—that contains 2,000 more schooling species; among them are the common fresh-water minnows, or shiners, and the familiar characins of the tabletop aquarium. It is evident that these fishes must have converged on the schooling way of life by diverse evolutionary pathways. Of the marine fishes the best-known schooling orders are three that rank among the most numerous in the sea and constitute a vast portion of the world's fish supply. They are the Clupeiformes, the well-known herrings; the Mugiliformes, which include, in addition to the schooling mullets and our laboratory silversides, the solitary barracuda; the Perciformes, comprising the schooling jacks, pompanos, bluefishes, mackerels and tuna and the occasionally schooling snappers and grunts as well as numerous families of nonschoolers. Anatomically the Clupeiformes and the Mugiliformes are rather primitive fishes, whereas the Perciformes are advanced.

Although unrelated, these fishes do have significant features in common. Like many other schooling fishes, they are generally sleek and silvery. Significantly, they also have the same small and flattened pectoral fins actuated by musculature that does not permit much mobility. As C. M. Breder, Jr., of the American Museum of Natural History was the first to observe, these fishes cannot swim backward. When they make a pass at a bit of food and happen to miss it, they must come around on a wide turn for another attempt. This limitation on their maneuverability must nonetheless be an advantage in the maintenance of a school, because it tends to keep them all moving forward.

Since the schooling families include anatomically primitive as well as advanced forms, the evidence from living species does not show whether schooling is a primitive or an advanced adaptation. The fossil record is equally inconclusive on this score. Herrings are found in great number in Eocene deposits and one may reasonably speculate that they were schooling then. But fishes were evolving long before the Eocene, and it is impossible to determine one way or the other whether the fishes of those times schooled.

In spite of all the indications that schooling is an effective adaptation, no student of the subject has been able to show why it is so effective. Many advantages can be cited in favor of the behavior, but none seems critical to survival. It is said, for example, that the school creates for its predators, as it does for human observers, the illusion that it is a huge and formidable animal of some kind and so frightens off the predator. No real evidence supports this idea, and one can more plausibly see the school as providing easy prey. If the predator misses one fish, there is always another. In an experiment with goldfish, on the other hand, Carl Welty of Beloit College found that the fish consumed fewer *Daphnia* when they

SCHOOL OF ATLANTIC MENHADEN in Long Island Sound was photographed from the air by Jan Hahn of the Woods Hole Oceanographic Institution. The menhaden, which is a species of herring, forms schools containing as many as a million members.

were fed too many than they did if they were allowed a smaller number. Welty suggested that large numbers of prey might "confuse" the predator. This idea finds support in a mathematical analysis by Vernon E. Brock and Robert H. Riffenburgh of the University of Hawaii, which shows that a school cannot be decimated by attackers once it exceeds a certain number. But one must then ask: Why do some predators school?

Another rationalization for schooling holds that it facilitates the finding of food. When it comes to the search itself, however, only the fish on the school's periphery will be in a position to locate the food; the talents of those in the center of the school are wasted. Of course, once the food is sighted, all may partake. The young of many fishes travel in schools, and their social feeding seemingly promotes more rapid growth. As our efforts to raise fry in isolation would indicate, the sight (or taste and smell) of other fish feeding induces fish to feed. Again, one must doubt that this advantage could account for the evolution of schooling behavior in so many different species.

Another advantage, often cited, has to do with the reproduction of the schooling species. When it is time to reproduce, there is no courtship behavior, no mate selection; as Parr observed some years ago, the males and females of schooling species are usually indistinguishable on casual inspection. The fishes simply shed their eggs and sperm in almost countless numbers into the plankton and leave the spawning site. This certainly enhances the probability of successful fertilization. In some of my collecting, however, I have found schools that were either all male or all female!

To the list of potential adaptive advantages I would like to add another one. Hydrodynamic considerations argue that schooling provides a more efficient way to move through the water. The exertion of each fish may be lessened because it can utilize the turbulence produced by the surrounding fish. Although the fish at the leading edge of the school may have to expend no less energy than solitary fish, the followers may receive enough assistance to help reduce their expenditure of energy. The attainment of maximum efficiency may dictate an optimum fish-to-fish distance in the school.

Study of the schooling of fishes has asked more questions than it has answered. But the questions have now begun to suggest fruitful programs of observation and experiment.

"Imprinting" in a Natural Laboratory

by Eckhard H. Hess
August 1972

*A synthesis of laboratory and field techniques has
led to some interesting discoveries about imprinting,
the process by which newly hatched birds rapidly
form a permanent bond to the parent.*

In a marsh on the Eastern Shore of Maryland, a few hundred feet from my laboratory building, a female wild mallard sits on a dozen infertile eggs. She has been incubating the eggs for almost four weeks. Periodically she hears the faint peeping sounds that are emitted by hatching mallard eggs, and she clucks softly in response. Since these eggs are infertile, however, they are not about to hatch and they do not emit peeping sounds. The sounds come from a small loudspeaker hidden in the nest under the eggs. The loudspeaker is connected to a microphone next to some hatching mallard eggs inside an incubator in my laboratory. The female mallard can hear any sounds coming from the laboratory eggs, and a microphone beside her relays the sounds she makes to a loudspeaker next to those eggs.

The reason for complicating the life of an expectant duck in such a way is to further our understanding of the phenomenon known as imprinting. It was through the work of the Austrian zoologist Konrad Z. Lorenz that imprinting became widely known. In the 1930's Lorenz observed that newly hatched goslings would follow him rather than their mother if the goslings saw him before they saw her. Since naturally reared geese show a strong attachment for their parent, Lorenz concluded that some animals have the capacity to learn rapidly and permanently at a very early age, and in particular to learn the characteristics of the parent. He called this process of acquiring an attachment to the parent *Prägung*, which in German means "stamping" or "coinage" but in English has been rendered as "imprinting." Lorenz regarded the phenomenon as being different from the usual kind of learning because of its rapidity and apparent permanence. In fact, he was hesitant at first to regard imprinting as a form of learn-

ing at all. Some child psychologists and some psychiatrists nevertheless perceived a similarity between the evidence of imprinting in animals and the early behavior of the human infant, and it is not surprising that interest in imprinting spread quickly.

From about the beginning of the 1950's many investigators have intensively studied imprinting in the laboratory. Unlike Lorenz, the majority of them have regarded imprinting as a form

of learning and have used methods much the same as those followed in the study of associative learning processes. In every case efforts were made to manipulate or stringently control the imprinting process. Usually the subjects are incubator-hatched birds that are reared in the laboratory. The birds are typically kept isolated until the time of the laboratory imprinting experience to prevent interaction of early social experience and the imprinting experience. Various objects

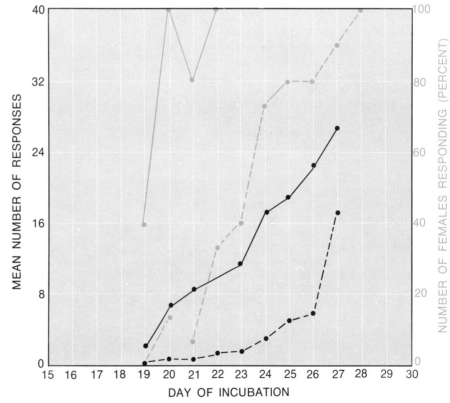

VOCAL RESPONSES to hatching-duckling sounds of 15 female wild mallards (*broken curves*) and five human-imprinted mallards (*solid curves*), which were later released to the wild, followed the same pattern, although the human-imprinted mallards began responding sooner and more frequently. A tape recording of the sounds of a hatching duckling was played daily throughout the incubation period to each female mallard while she was on her nest. Responses began on the 19th day of incubation and rose steadily until hatching.

have been used as artificial parents: duck decoys, stuffed hens, dolls, milk bottles, toilet floats, boxes, balls, flashing lights and rotating disks. Several investigators have constructed an automatic imprinting apparatus into which the newly hatched bird can be put. In this kind of work the investigator does not observe the young bird directly; all the bird's movements with respect to the imprinting object are recorded automatically.

Much of my own research during the past two decades has not differed substantially from this approach. The birds I have used for laboratory imprinting studies have all been incubated, hatched and reared without the normal social and environmental conditions and have then been tested in an artificial situation. It is therefore possible that the behavior observed under such conditions is not relevant to what actually happens in nature.

It is perhaps not surprising that studies of "unnatural" imprinting have produced conflicting results. Lorenz' original statements on the permanence of natural imprinting have been disputed. In many instances laboratory imprinting experiences do not produce permanent and exclusive attachment to the object selected as an artificial parent. For example, a duckling can spend a considerable amount of time following the object to which it is to be imprinted, and immediately after the experience it will follow a completely different object.

In one experiment in our laboratory we attempted to imprint ducklings to ourselves, as Lorenz did. For 20 continuous hours newly hatched ducklings were exposed to us. Before long they followed us whenever we moved about. Then they were given to a female mallard that had hatched a clutch of ducklings several hours before. After only an hour and a half of exposure to the female mallard and other ducklings the human-imprinted ducklings followed the female on the first exodus from the nest. Weeks later the behavior of the human-imprinted ducks was no different from the behavior of the ducks that had been hatched in the nest. Clearly laboratory imprinting is reversible.

We also took wild ducklings from their natural mother 16 hours after hatching and tried to imprint them to humans. On the first day we spent many hours with the ducklings, and during the next two months we made lengthy attempts every day to overcome the ducklings' fear of us. We finally gave up. From the beginning to the end the ducks

remained wild and afraid. They were released, and when they had matured, they were observed to be as wary of humans as normal wild ducks are. This result suggests that natural imprinting, unlike artificial laboratory imprinting, is permanent and irreversible. I have had to conclude that the usual laboratory imprinting has only a limited resemblance to natural imprinting.

It seems obvious that if the effects of natural imprinting are to be understood, the phenomenon must be studied as it operates in nature. The value of such studies was stressed as long ago as 1914 by the pioneer American psychologist John B. Watson. He emphasized that field observations must always be made to test whether or not conclusions drawn from laboratory studies conform to what actually happens in nature. The disparity between laboratory results and what happens in nature often arises from the failure of the investigator to really look at the animal's behavior. For years I have cautioned my students against shutting their experimental animals in "black boxes" with automatic recording

devices and never directly observing how the animals behave.

This does not mean that objective laboratory methods for studying the behavior of animals must be abandoned. With laboratory investigations large strides have been made in the development of instruments for the recording of behavior. In the study of imprinting it is not necessary to revert to imprecise naturalistic observations in the field. We can now go far beyond the limitations of traditional field studies. It is possible to set up modern laboratory equipment in actual field conditions and in ways that do not disturb or interact with the behavior being studied, in other words, to achieve a synthesis of laboratory and field techniques.

The first step in the field-laboratory method is to observe and record the undisturbed natural behavior of the animal in the situation being studied. In our work on imprinting we photographed the behavior of the female mallard during incubation and hatching. We photographed the behavior of the ducklings during and after hatching. We recorded

CLUCKS emitted by a female wild mallard in the fourth week of incubating eggs are shown in the sound spectrogram (*upper illustration*). Each cluck lasts for about 150 milliseconds

all sounds from the nest before and after hatching. Other factors, such as air temperature and nest temperature, were also recorded.

A detailed inventory of the actual events in natural imprinting is essential for providing a reference point in the assessment of experimental manipulations of the imprinting process. That is, the undisturbed natural imprinting events form the control situation for assessing the effects of the experimental manipulations. This is quite different from the "controlled" laboratory setting, in which the ducklings are reared in isolation and then tested in unnatural conditions. The controlled laboratory study not only introduces new variables (environmental and social deprivation) into the imprinting situation but also it can prevent the investigator from observing factors that are relevant in wild conditions.

My Maryland research station is well suited for the study of natural imprinting in ducks. The station, near a national game refuge, has 250 acres of marsh and forest on a peninsula on which there are many wild and semiwild mallards. Through the sharp eyes of my technical assistant Elihu Abbott, a native of the Eastern Shore, I have learned to see much I might otherwise have missed. Initially we looked at and listened to the undisturbed parent-offspring interaction of female mallards that hatched their own eggs both in nests on the ground and in specially constructed nest boxes. From our records we noticed that the incubation time required for different clutches of eggs decreased progressively between March and June. Both the average air temperature and the number of daylight hours increase during those months; both are correlated with the incubation time of mallard eggs. It is likely, however, that temperature rather than photoperiod directly influences the duration of incubation. In one experiment mallard eggs from an incubator were slowly cooled for two hours a day in a room with a temperature of seven degrees Celsius, and another set of eggs was cooled in a room at 27 degrees C. These temperatures respectively correspond to the mean noon temperatures at the research station in March and in June. The eggs that were placed in the cooler room took longer to hatch, indicating that temperature affects the incubation time directly. Factors such as humidity and barometric pressure may also play a role.

We noticed that all the eggs in a wild nest usually hatch between three and eight hours of one another. As a result all the ducklings in the same clutch are approximately the same age in terms of the number of hours since hatching. Yet when mallard eggs are placed in a mechanical incubator, they will hatch over a two- or three-day period even when precautions are taken to ensure that all the eggs begin developing simultaneously. The synchronous hatching observed in nature obviously has some survival value. At the time of the exodus from the nest, which usually takes place between 16 and 32 hours after hatching, all the ducklings would be of a similar age and thus would have equal motor capabilities and similar social experiences.

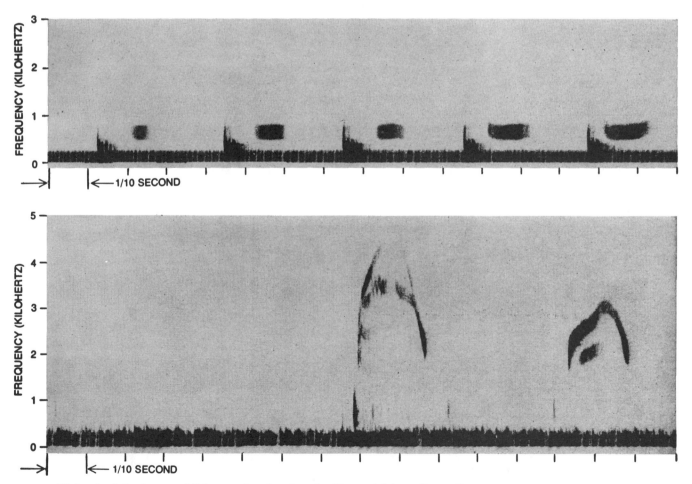

and is low in pitch: about one kilohertz or less. Sounds emitted by ducklings inside the eggs are high-pitched, rising to about four kilohertz (*lower illustration*). Records of natural, undisturbed imprinting events in the nest provide a control for later experiments.

Over the years our laboratory studies and actual observations of how a female mallard interacts with her offspring have pointed to the conclusion that imprinting is related to the age after hatching rather than the age from the beginning of incubation. Many other workers, however, have accepted the claim that age from the beginning of incubation determines the critical period for maximum effectiveness of imprinting. They base their belief on the findings of Gilbert Gottlieb of the Dorothea Dix Hospital in Raleigh, N.C., who in a 1961 paper described experiments that apparently showed that maximum imprinting in ducklings occurs in the period between 27 and 27½ days after the beginning of incubation. To make sure that all the eggs he was working with started incubation at the same time he first chilled the eggs so that any partially developed embryos would be killed. Yet the 27th day after the beginning of incubation can hardly be the period of maximum imprinting for wild ducklings that hatch in March under natural conditions, because such ducklings take on the average 28 days to hatch. Moreover, if the age of a duckling is measured from the beginning of incubation, it is hard to explain why eggs laid at different times in a hot month in the same nest will hatch within six to eight hours of one another under natural conditions.

Periodic cooling of the eggs seems to affect the synchronization of hatching. The mallard eggs from an incubator that were placed in a room at seven degrees C. hatched over a period of a day and a half, whereas eggs placed in the room at 27 degrees hatched over a period of two

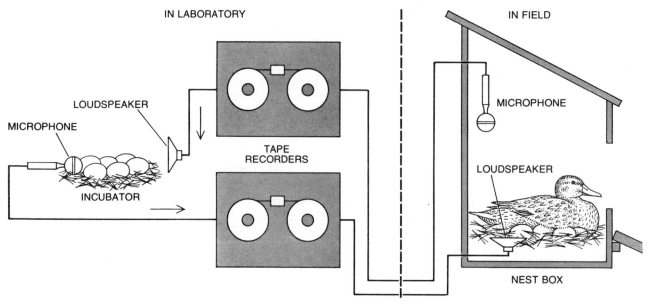

FEMALE MALLARD sitting on infertile eggs hears sounds transmitted from mallard eggs in a laboratory incubator. Any sounds she makes are transmitted to a loudspeaker beside the eggs in the laboratory. Such a combination of field and laboratory techniques permits recording of events without disturbing the nesting mallard and provides the hatching eggs with nearly natural conditions.

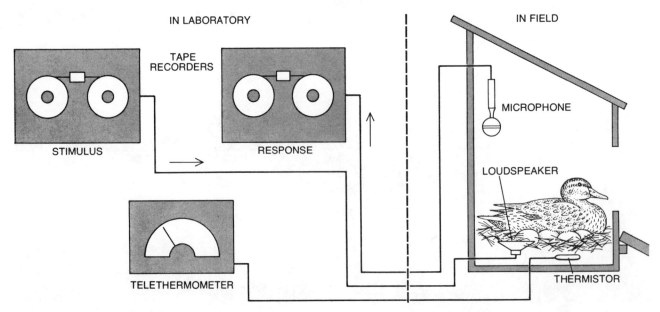

REMOTE MANIPULATION of prehatching sounds is accomplished by placing a sensitive microphone and a loudspeaker in the nest of a female wild mallard who is sitting on her own eggs. Prerecorded hatching-duckling sounds are played at specified times through the loudspeaker and the female mallard's responses to this stimulus are recorded. A thermistor probe transmits the temperature in the nest to a telethermometer and chart recorder. The thermistor records provide data about when females are on nest.

and a half days (which is about normal for artificially incubated eggs). Cooling cannot, however, play a major role. In June the temperature in the outdoor nest boxes averages close to the normal brooding temperature while the female mallard is absent. Therefore an egg laid on June 1 has a head start in incubation over those laid a week later. Yet we have observed that all the eggs in clutches laid in June hatch in a period lasting between six and eight hours.

We found another clue to how the synchronization of hatching may be achieved in the vocalization pattern of the brooding female mallard. As many others have noted, the female mallard vocalizes regularly as she sits on her eggs during the latter part of the incubation period. It seemed possible that she was vocalizing to the eggs, perhaps in response to sounds from the eggs themselves. Other workers had observed that ducklings make sounds before they hatch, and the prehatching behavior of ducklings in response to maternal calls has been extensively reported by Gottlieb.

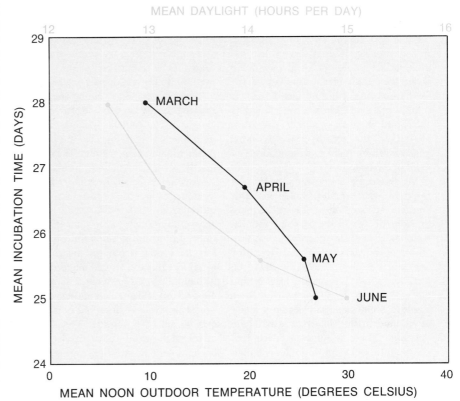

INCUBATION TIME of mallard eggs hatched naturally in a feral setting at Lake Cove, Md., decreased steadily from March to June. The incubation period correlated with both the outdoor temperature (*black curve*) and the daily photoperiod (*colored curve*).

We placed a highly sensitive microphone next to some mallard eggs that were nearly ready to hatch. We found that the ducklings indeed make sounds while they are still inside the egg. We made a one-minute tape recording of the sounds emitted by a duckling that had pipped its shell and was going to hatch within the next few hours. Then we made a seven-minute recording that would enable us to play the duckling sounds three times for one minute interspersed with one-minute silences. We played the recording once each to 37 female mallards at various stages of

NEST EXODUS takes place about 16 to 32 hours after hatching. The female mallard begins to make about 40 to 65 calls per minute and continues while the ducklings leave the nest to follow her. The ducklings are capable of walking and swimming from hatching.

incubation. There were no positive responses from the female mallards during the first and second week of incubation. In fact, during the first days of incubation some female mallards responded with threat behavior: a fluffing of the feathers and a panting sound. In the third week some females responded to the recorded duckling sounds with a few clucks. In the fourth week maternal clucks were frequent and were observed in all ducks tested.

We found the same general pattern of response whether the female mallards were tested once or, as in a subsequent experiment, tested daily during incubation. Mallards sitting on infertile eggs responded just as much to the recorded duckling sounds as mallards sitting on fertile eggs did. Apparently after sitting on a clutch of eggs for two or three weeks a female mallard becomes ready to respond to the sounds of a hatching duckling. There is some evidence that the parental behavior of the female mallard is primed by certain neuroendocrine mechanisms. We have begun a study of the neuroendocrine changes that might accompany imprinting and filial behavior in mallards.

To what extent do unhatched ducklings respond to the vocalization of the female mallard? In order to find out we played a recording of a female mallard's vocalizations to ducklings in eggs that had just been pipped and were scheduled to hatch within the next 24 hours. As before, the sounds were interspersed with periods of silence. We then recorded all the sounds made by the ducklings during the recorded female mallard vocalizations and also during the silent periods on the tape. Twenty-four hours before the scheduled hatching the ducklings emitted 34 percent of their sounds during the silent periods, which suggests that at this stage they initiate most of the auditory interaction. As hatching time approaches the ducklings emit fewer and fewer sounds during the silent periods. The total number of sounds they make, however, increases steadily. At the time of hatching only 9 percent of the sounds they make are emitted during the silent periods. One hour after hatching, in response to the same type of recording, the ducklings gave 37 percent of their vocalizations during the silent periods, a level similar to the level at 24 hours before hatching.

During the hatching period, which lasts about an hour, the female mallard generally vocalizes at the rate of from zero to four calls per one-minute inter-

val. Occasionally there is an interval in which she emits as many as 10 calls. When the duckling actually hatches, the female mallard's vocalization increases dramatically to between 45 and 68 calls per minute for one or two minutes.

Thus the sounds made by the female mallard and by her offspring are complementary. The female mallard vocalizes most when a duckling has just hatched. A hatching duckling emits its cries primarily when the female is vocalizing.

After all the ducklings have hatched the female mallard tends to be relatively quiet for long intervals, giving between zero and four calls per minute. This continues for 16 to 32 hours until it is time for the exodus from the nest. As the exodus begins the female mallard quickly builds up to a crescendo of between 40 and 65 calls per minute; on rare occasions we have observed between 70 and 95 calls per minute. The duration of the high-calling-rate period depends on how quickly the ducklings leave the nest to follow her. There is now a change in the sounds made by the female mal-

lard. Up to this point she has been making clucking sounds. By the time the exodus from the nest takes place some of her sounds are more like quacks.

The auditory interaction of the female mallard and the duckling can begin well before the hatching period. As I have indicated, the female mallard responds to unhatched-duckling sounds during the third and fourth week of incubation. Normally ducklings penetrate a membrane to reach an air space inside the eggshell two days before hatching. We have not found any female mallard that vocalized to her clutch before the duckling in the egg reached the air space. We have found that as soon as the duckling penetrates the air space the female begins to cluck at a rate of between zero and four times per minute. Typically she continues to vocalize at this rate until the ducklings begin to pip their eggs (which is about 24 hours after they have entered the air space). As the eggs are being pipped the female clucks at the rate of between 10 and 15 times per minute. When the pipping is completed, she

SOUND SPECTROGRAM of the calls of newly hatched ducklings in the nest and the mother's responses is shown at right. The high-pitched peeps of the ducklings are in the

DISTRESS CALLS of ducklings in the nest evoke a quacklike response from the female mallard. The cessation of the distress calls and the onset of normal duckling peeping sounds

drops back to between zero and four calls per minute. In the next 24 hours there is a great deal of auditory interaction between the female and her unhatched offspring; this intense interaction may facilitate the rapid formation of the filial bond after hatching, although it is quite possible that synchrony of hatching is the main effect. Already we have found that a combination of cooling the eggs daily, placing them together so that they touch one another and transmitting parent-young vocal responses through the microphone-loudspeaker hookup between the female's nest and the laboratory incubator causes the eggs in the incubator to hatch as synchronously as eggs in nature do. In fact, the two times we did this we found that all the eggs in the clutches hatched within four hours of one another. It has been shown in many studies of imprinting, including laboratory studies, that auditory stimuli have an important effect on the development of filial attachment. Auditory stimulation, before and after hatching, together with tactile

stimulation in the nest after hatching results in ducklings that are thoroughly imprinted to the female mallard that is present.

Furthermore, it appears that auditory interaction before hatching may play an important role in promoting the synchronization of hatching. As our experiments showed, not only does the female mallard respond to sounds from her eggs but also the ducklings respond to her clucks. Perhaps the daily cooling of the eggs when the female mallard leaves the nest to feed serves to broadly synchronize embryonic and behavioral development, whereas the auditory interaction of the mother with the ducklings and of one duckling with another serves to provide finer synchronization. Margaret Vince of the University of Cambridge has shown that the synchronization of hatching in quail is promoted by the mutual auditory interaction of the young birds in the eggs.

Listening to the female mallards vocalize to their eggs or to their newly hatched offspring, we were struck by the

fact that we could tell which mallard was vocalizing, even when we could not see her. Some female mallards regularly emit single clucks at one-second intervals, some cluck in triple or quadruple clusters and others cluck in clusters of different lengths. The individual differences in the vocalization styles of female mallards may enable young ducklings to identify their mother. We can also speculate that the characteristics of a female mallard's voice are learned by her female offspring, which may then adopt a similar style when they are hatching eggs of their own.

The female mallards not only differ from one another in vocalization styles but also emit different calls in different situations. We have recorded variations in pitch and duration from the same mallard in various nesting situations. It seems likely that such variations in the female mallard call are an important factor in the imprinting process.

Studies of imprinting in the laboratory have shown that the more effort a duckling has to expend in following the im-

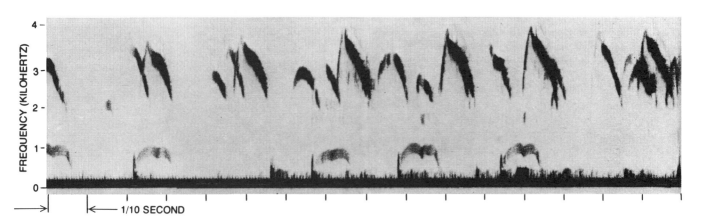

two-to-four-kilohertz range. They normally have the shape of an inverted *V*. The female mallard's clucks are about one kilohertz

and last about 130 milliseconds. After the eggs hatch the vocalization of the female changes both in quantity and in quality of sound.

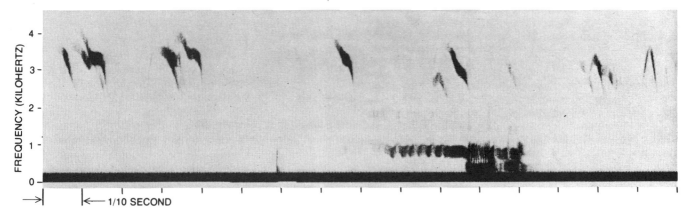

is almost immediate, as can be seen in this sound spectrogram. The female mallard's quacklike call is about one kilohertz in pitch and

has a duration of approximately 450 milliseconds. The call is emitted about once every two seconds in response to distress cries.

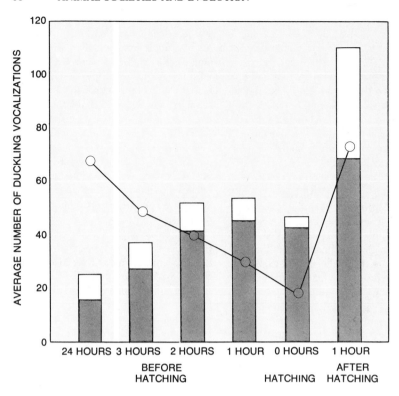

NUMBER OF SOUNDS from ducklings before and after hatching are shown. The ducklings heard a recording consisting of five one-minute segments of a female mallard's clucking sounds interspersed with five one-minute segments of silence. The recording was
played to six mallard eggs and the number of vocal responses by the ducklings to the
clucking segments (gray bars) and to the silent segments (white bars) were counted.
Twenty-four hours before hatching 34 percent of the duckling sounds were made during
the silent interval, indicating the ducklings initiated a substantial portion of the early auditory interaction. As hatching time approached the ducklings initiated fewer and fewer of
the sounds and at hatching vocalized most in response to the clucks of the female mallard.

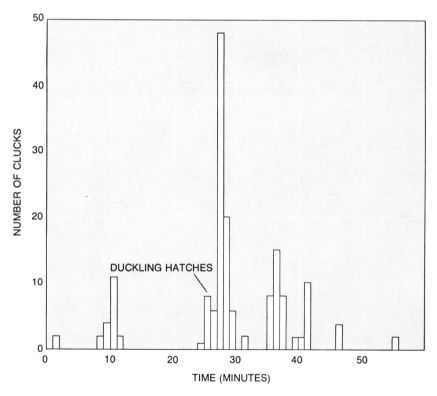

CLUCKING RATE of a wild, ground-nesting female mallard rose dramatically for about
two minutes while a duckling hatched and then slowly declined to the prehatching rate.
Each bar depicts the number of clucks emitted by the female during a one-minute period.

printing object, the more strongly it prefers that object in later testing. At first
it would seem that this is not the case
in natural imprinting; young ducklings
raised by their mother have little difficulty following her during the exodus
from the nest. Closer observation of
many nests over several seasons showed,
however, that ducklings make a considerable effort to be near their parent.
They may suffer for such efforts, since
they can be accidentally stepped on,
squeezed or scratched by the female
adult. The combination of effort and
punishment may actually strengthen imprinting. Work in my laboratory showed
that chicks given an electric shock while
they were following the imprinting object later showed stronger attachment to
the object than unshocked chicks did.
It is reasonable to expect similar results
with ducklings.

Slobodan Petrovich of the University
of Maryland (Baltimore County) and I
have begun a study to determine the relative contributions of prehatching and
posthatching auditory experience on imprinting and filial attachment. The auditory stimuli consist of either natural mallard maternal clucks or a human voice
saying "Come, come, come." Our results
indicate that prehatching stimulation by
natural maternal clucks may to a degree
facilitate the later recognition of the
characteristic call of the mallard. Ducklings lacking any experience with a maternal call imprint as well to a duck decoy that utters "Come, come, come" as
to a decoy that emits normal mallard
clucks. Ducklings that had been exposed
to a maternal call before hatching imprinted better to decoys that emitted the
mallard clucks. We found, however, that
the immediate posthatching experiences,
in this case with a female mallard on the
nest, can highly determine the degree of
filial attachment and make imprinting
to a human sound virtually impossible.

It is important to recognize that almost
all laboratory imprinting experiments,
including my own, have been deprivation experiments. The justification for
such experiments has been the ostensible
need for controlling the variables of the
phenomenon, but the deprivation may
have interfered with the normal behavioral development of the young ducklings. Whatever imprinting experiences
the experimenter allows therefore do not
produce the maximum effect.

Although our findings are far from
complete, we have already determined
enough to demonstrate the great value of
studying imprinting under natural conditions. The natural laboratory can be

profitably used to study questions about imprinting that have been raised but not answered by traditional laboratory experiments. We must move away from the in vitro, or test-tube, approach to the study of behavior and move toward the in vivo method that allows interaction with normal environmental factors. Some of the questions are: What is the optimal age for imprinting? How long must the imprinting experience last for it to have the maximum effect? Which has the greater effect on behavior: first experience or the most recent experience? Whatever kind of behavior is being studied, the most fruitful approach may well be to study the behavior in its natural context.

8

The Social Ecology of Coyotes

by Marc Bekoff and Michael C. Wells
April 1980

The nature of their food supply seems to determine whether they live alone or in a pack. Such patterns of behavior may bear on the question of whether or not they are a threat to livestock.

Motion-picture films about the American West almost always depict coyotes in the same way, as solitary animals howling mournfully on the top of a distant hill. In reality coyotes are protean creatures that display a wide range of behavior. They are characterized by highly variable modes of social organization, ranging from solitary (except for the breeding season) and transient individuals to gregarious and stable groups that may live in the same area over a long period of time. Between the two extremes are single individuals and mated pairs that tend to remain in one area. Indeed, a single coyote may in its lifetime experience all the different grades of sociality. This remarkable flexibility in the ways coyotes interact with one another can best be understood by examining their ecology, or the ways they interact with their environment.

It is generally accepted that most animal characteristics are the product of an interaction between inherited predispositions and the environment. In other words, although the cumulative passing of genes by successfully reproducing individuals establishes certain tendencies in each animal, many observable traits are subject to modification by proximate, or immediate, factors in the animal's environment. Thus many of an animal's traits, in particular behavioral ones, can be viewed as adaptations to the environments in which the animal has lived or is living. For example, the Dutch ethologist Hans Kruuk, who has done intensive studies of hyenas, has concluded that for many large carnivores, which typically have few predators (other than man), the nature of food resources is an important proximate factor that influences social behavior. More precisely, it appears that variations in the sociality of carnivores of the same species can often be traced to differences in their food supply.

For the past three years we have been observing the behavior of coyotes in the wild, mostly in Grand Teton National Park near the town of Jackson in northwestern Wyoming. Our studies indicate that the social organization of coyotes is indeed a reflection of their food resources and that three variables have a direct and significant impact in this regard: the size of the available prey, the prey's spatial distribution and its temporal, or seasonal, distribution. We shall report on our findings about both the specific behavioral adaptations coyotes seem to make to different types of food supply and the advantages these adaptations seem to confer. Before we undertake to sort out this aspect of the complex relation between coyotes and their environment we shall briefly describe the animals and the setting in which we are studying them.

Coyotes (*Canis latrans*) belong to the same mammalian family as jackals, foxes, wolves and domestic dogs. There are 19 recognized subspecies of coyotes, but because the animals are currently more mobile than they used to be and crossbreed to a greater extent there seems little reason to retain the more refined classification. Coyotes mate once a year and are generally monogamous, so that the same pair may mate in the same area over long periods, often returning to the same den site year after year. (Coyotes bear their young in holes in the ground, which they may or may not dig for themselves; the coyotes we observed generally made use of holes that had already been excavated by badgers.)

In a study of coyotes in the Canadian province of Alberta, Donald Bowen of the University of British Columbia noted that coyotes living in packs not only eat, sleep and travel in close association with one another but also tend to exhibit dominance relations. Franz J. Camenzind, who has studied coyotes on the National Elk Refuge adjoining the town of Jackson, has made similar observations. In general pack members are more sociable with one another than they are with outsiders, such as single coyotes living in the same area or passing through. It appears that most members of a coyote pack are genetically related. Indeed, the basis of coyote so-

cial structure is probably the mated pair supplemented by those offspring that do not leave the pack when they are old enough to care for themselves.

Typically only one male and one female breed in each pack. Some of the nonbreeding individuals may help to raise other members of the pack, most probably their younger siblings, and to defend food supplies, mainly against other coyotes. Packs may also include nonbreeding hangers-on, probably also offspring of the mated pair in the pack, that continue to live in the vicinity of the pack but interact very little with it. (It is possible that these individuals benefit from such a minimal association by "inheriting" a breeding area after a parent leaves it or dies.)

Coyotes are found in diverse habitats in Canada, Central America and most of the states of the continental U.S., but even within a single geographical setting their social behavior can vary dramatically. Our primary site for the long-term observation of wild coyotes is the area around Blacktail Butte in the southeastern corner of Grand Teton National Park. This is a particularly good place for a study of behavior and ecology because the animals that live in the park are relatively unaffected by man. Moreover, from Blacktail Butte, which rises some 300 meters from the surrounding valley floor, it is easy to observe coyotes going about their normal activities. Our findings for the Blacktail Butte area have been supplemented by observations of coyotes one of us (Bekoff) made with the aid of several students in Rocky Mountain National Park in Colorado. In the Moraine Park section of that park, where the study was carried out, the environmental conditions were quite different from those found at Blacktail Butte, and so in many cases comparing data from the two locations has helped us to identify variables influencing social behavior. We have also done experiments with animals in captivity, so that relevant competing variables could be more closely controlled.

For studies such as ours it is important to be able to identify various members of a wild population, but in the case of coyotes distinguishing characteristics such as size (ranging from eight to 20 kilograms for males) and coat color (a highly variable blend of white, gray, brown and rust) may change with time. As a result it has been necessary to capture and mark individual coyotes, and for this purpose we generally rely on foot traps, the jaws of which are wrapped with thick cotton padding to reduce the likelihood of injury to the trapped animal. To keep the coyote from thrashing around in the trap we frequently attach a tranquilizer pellet, which the animal usually swallows. The tranquilizer sedates the trapped coyote but does not render it unconscious. The trap lines are covered on foot, on skis or by automobile every six to eight hours so that the coyotes are restrained no longer than is necessary.

Once a coyote has been captured it becomes extremely docile, and so when we find a coyote in one of our traps, we immediately release it and then proceed to weigh it, note its sex, make an assessment of its physical condition and estimate its age. Next we attach a colored identification tag to each ear and fit it with a collar bearing a small radio trans-

mitter. In this way after the coyote is released we can identify it even when it is out of sight, and we can always tell which coyotes are associating with one another. Because the area around Blacktail Butte is quite open, however, we are usually able to see the coyotes (with binoculars or a spotting telescope if not with the eye), and the radio transmitters serve primarily for the gathering of data on wide-ranging movements of individuals and groups.

There are many ways in which the nature of food resources might influence the social behavior of coyotes. For example, when large prey animals such as ungulates (hoofed mammals) are available, several carnivores (including lions, wolves, jackals and African wild dogs) have been seen to band together in packs for cooperative hunting. Pack living may also be an adaptation for the defense of major food supplies such as caches of carrion. The observations of David Macdonald of the Animal Behavior Research Group at the University of Oxford indicate that this is the case for golden jackals (*Canis aureus*) found in Israel. We have observed that for coyotes, at least in the conditions under

which we are observing them, group hunting is a rare and generally unsuccessful undertaking. In fact, from our vantage on Blacktail Butte we have never seen either a group of coyotes or a single coyote attacking a large live ungulate. On the other hand, our findings and Bowen's indicate that coyotes do group together to defend certain food resources.

In the area around Blacktail Butte there is a significant seasonal fluctuation in the food items that sustain coyotes. In "summer" (the period from May through October) the coyotes feed mainly on rodents such as pocket gophers, field mice and Uinta ground squirrels. In "winter" (the period from November through April) the major food supply is the carrion of ungulates such as deer, moose and in particular elk that have died from causes other than coyote predation. To put it another way, in summer the coyotes hunt and kill small prey that are generally distributed widely over the area in which the coyotes live and in winter they feed on large dead prey (mainly elk) that because of the formation of herds and legal hunting by human beings during a limited season generally tend to be distributed

MEMBERS OF A COYOTE PACK gather around the carcass of an elk in the snow on the National Elk Refuge adjoining the town of Jackson, Wyo. Coyotes display remarkably flexible patterns of social organization, ranging from transient individuals and mated pairs to large, stable groups that tend to remain in one area. Studies of these animals in the wild indicate that pack living represents an adaptation to large, clumped food resources such as the carrion of ungulates (hoofed mammals), whereas solitary living is associated with the availability of small live prey such as rodents (*see illustration on page 73*). Coyotes have rarely been observed to prey on the large ungulates (elk, moose and so on) whose carrion generally sustains them in winter; the elk shown in this photograph died of other causes.

as isolated clumps of carrion. The increased availability of carrion in winter is a widespread phenomenon, largely as a result of the higher ungulate mortality in that season.

Our basic hypotheses about the role that the size of food items and their spatial and seasonal distribution play in molding coyotes' social behavior suggest that it should be possible to see variations in the sociality not only of populations of coyotes with access to different food resources but also within a single population from season to season.

To determine the effects of the seasonal fluctuation of prey at Blacktail Butte we compared the sizes of the coyote groups we found there in summer and in winter. Between September, 1977, and August, 1979, we made more than 1,000 sightings of 35 marked coyotes and about 15

PACK MEMBER DEFENDS CARRION (not visible) by "threat gaping" at an intruding coyote to chase it away. The two coyotes at the lower right belong to the same pack as the coyote at the center does and so remain unthreatened near the carrion. Ability to successfully defend such a food supply appears to be one of principal advantages of pack living. Photograph was taken by Franz J. Camenzind.

THREE-WEEK-OLD COYOTE PUPS require feeding and protection and remain close to the hole in the ground that serves as their den. (Although coyotes may excavate their own den, these pups are in an abandoned badger hole their parents enlarged.) Coyote pups begin to make forays away from the den when they are two to three months old, and they may strike out on their own when they are six to nine months old. It seems that most members of a coyote pack are genetically related and that the basis of coyote social structure is probably the mated pair supplemented by a number of nondispersing offspring. In most instances only one male-female pair in a pack breeds.

unmarked ones and found that in the summer months, when rodents were the major food resource, the average group size was 1.3 individuals and that in winter the average rose to 1.8. Hence the availability of large, clumped prey items did seem to be correlated with heightened sociability in these coyotes.

Moreover, we made another interesting discovery when we compared our findings with Camenzind's for coyotes on the National Elk Refuge. Camenzind's observation site is only about seven kilometers from our own, but since many more elk winter there, the supply of ungulate carrion is larger and denser. Camenzind found that on the elk refuge the coyote groups were also larger, with an average group size of 1.6 individuals in summer and three in winter. This finding suggests that the increased availability of ungulate carrion in winter not only serves to increase sociability in that season but also may have a cumulative effect, resulting in increased gregariousness the following summer. It is also interesting to note that in the Moraine Park area of Rocky Mountain National Park, where for three successive winters there was virtually no ungulate carrion, the situation was quite different. The coyotes were forced to depend on small rodents throughout the year, and the average group size in both summer and winter was 1.1.

We also compared the frequency with which three coyote social groupings—single individuals, mated pairs and packs of three or more individuals—were sighted at the various observation areas over an entire year. For example, at Blacktail Butte 35 percent of our sightings were of packs and about 50 percent were of single individuals, either transients passing through an area occupied by a pack (or by a mated pair) or solitary coyotes living on the edges of the area. On the carrion-rich elk refuge, however, only about 15 percent of Camenzind's observations were of single coyotes and about 60 percent were of packs. It would appear that in the vicinity of Blacktail Butte, where ungulate carrion is scarcer and is clumped in only a few small areas, fewer individuals can live in packs that defend these resources. The remaining coyotes, which are generally excluded from the clumped carrion, must forage widely for food, either alone or as a mated pair. This conclusion is supported by the fact that at the Rocky Mountain National Park site, where there was almost no carrion, 97 percent of the sightings were of single individuals.

In order to gain a better understanding of the nature of coyote groupings and the advantages of the adaptation to defendable resources we did not have to cover a large area. Indeed, the observation over the past three years of two groups of coyotes with contiguous home

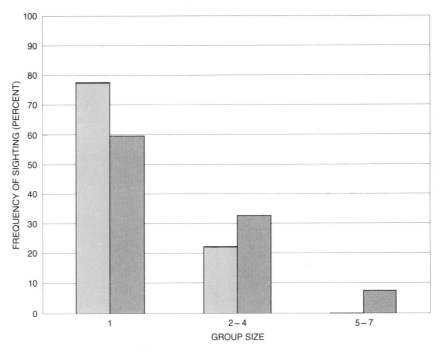

SEASONAL VARIATION in the sociability of coyotes may reflect a seasonal fluctuation in the availability of different prey items. This chart compares the frequency with which groups of different sizes were sighted from Blacktail Butte in Grand Teton National Park in Wyoming during the "summer" season (May through October) and the "winter" season (November through April). In summer (*color*), when coyotes sustained themselves by catching rodents, they were significantly less social than in winter (*light gray*), when ungulate carrion was available.

ranges in the vicinity of Blacktail Butte has provided us with ample evidence of the ways in which food supply can influence social behavior. (An animal's home range is defined as the area it covers routinely in the course of its daily activities.)

For example, in the winter of 1978–79 there was a significant difference in the quantity of elk carrion found on the two home ranges. Completely by chance (no attempt was made to control the distribution of carrion in the Blacktail Butte area) Group A had about 17 percent of the available carrion and Group B about 83 percent. As might be expected, Group A was the smaller one, consisting from November, 1978, through April, 1979, of only a single mated pair. All the young of the pair from previous years had dispersed. In the same period Group B had four members: a mated pair, an adult male born to them in 1977 and a male yearling born to them in 1978. (The older nonbreeding male helped to raise its siblings born in 1978.) The group also included two female hangers-on, one that was born to the mated pair in 1978 and another that we believe was born to them in 1977; these individuals rarely interacted with the members of the pack but were allowed to remain in their vicinity. From November, 1978, through the following May (and beyond) the four main members of Group B were highly cohesive: eating, sleeping, traveling and defending carrion in close association with one another. In this period only 6 percent of

the sightings of pack members were of single individuals and more than 50 percent were of all four pack members together. From November through April the male and female of Group A were observed together 71 percent of the time, and on the remaining occasions each animal was seen in the vicinity of other coyotes, although not in close association with them.

It has been observed that when coyotes other than a mated pair spend a winter together, there is an increased probability they will also spend the summer together. Our observations of the two groups in the area of Blacktail Butte indicate that when winter food is in good supply, older pups may continue to share at least a part of their parents' home range, and that if the pups remain in association with their parents throughout their first winter, there is a good chance that as yearlings they will remain through the following summer and perhaps beyond. It is interesting to note that two of the young that left the home range of Group A (the mated pair) in the fall of 1978 returned to it (from the National Elk Refuge, where they had spent the winter) the following spring, their return coinciding with the seasonal increase in rodents on the parental home range. These yearlings have remained solitary, not helping to raise their younger siblings, and in general they appear to be less closely bonded to their parents than the yearlings in Group B, which never left the pack.

During the past winter (1979–80)

there has been another interesting development in the relation between food availability and social organization in the coyote groups living in the vicinity of Blacktail Butte. In the previous two winters heavy snows fell in our study area in December, but this year snow did not blanket the home ranges of Group *A* (the mated pair) and Group *B* (the pack) until late January. As a result rodents were available in greater number and for a longer period than they had been in the preceding winters, supplementing the usual winter supply of elk carrion. In the previous two winters all the young from Group *A* had dispersed by November, but this year a juvenile born in April, 1979, was still with its parents in February. (In Group *B* three juveniles born in April, 1979, still remained with the pack in February.) Thus it appears that a naturally occurring change in the coyotes' food resources resulted in a change in their social organization, at least over a short period of time. The consequences of this change will be investigated in the future.

Social bonding is not the only aspect of coyotes' social behavior that is affected by variations in the food supply. Such variations also have a strong influence on how the animals make use of space. For the purposes of this discussion it is important to understand the distinction between a home range, the area an animal or a group of animals covers routinely in the course of its daily activities, and a territory. A home range has a flexible, undefended boundary, so that the home ranges of different individuals or groups may overlap considerably. A territory, on the other hand, is defined as the area that an individual or group occupies to the almost complete exclusion of other animals of the same species and that it will actively defend against them. In some geographical areas coyotes clearly defend their territory against other animals, but in other areas there is no evidence that they are territorial. Our own findings indicate it is only coyotes in packs that are territorial; individuals with a fixed home range but living alone or in mated pairs are not. Consider the two coyote groups we observed in the area of Blacktail Butte.

The four members of Group *B* maintained as a group a territory with rigorous boundaries between themselves and Group *A*, the mated pair. They also repelled many other coyotes from their territory, sometimes chasing an intruder for as much as two or three kilometers. (In April, 1979, we saw the breeding female of the pack chase an intruding coyote for a kilometer only a few days after she had given birth to a litter, and when she returned to the den, her mate chased the intruder for three more kilometers.) On the other hand, the two members of Group *A* were never seen defending a part of their home range against any

other coyote. These findings, which are confirmed by those of other workers, indicate that the intensity with which an area is defended by individuals or groups is related to the presence of a large, clumped food resource.

We also found that a shortage of food clearly brings about increased trespassing into neighboring home ranges and territories, particularly those in which desired food items can be found. For example, although Group *A*, the mated pair, made frequent forays into the territory defended by Group *B*, no member of Group *B* was ever observed intruding onto the home range of Group *A*. In fact, the members of the pack rarely left their own territory, which is not surpris-

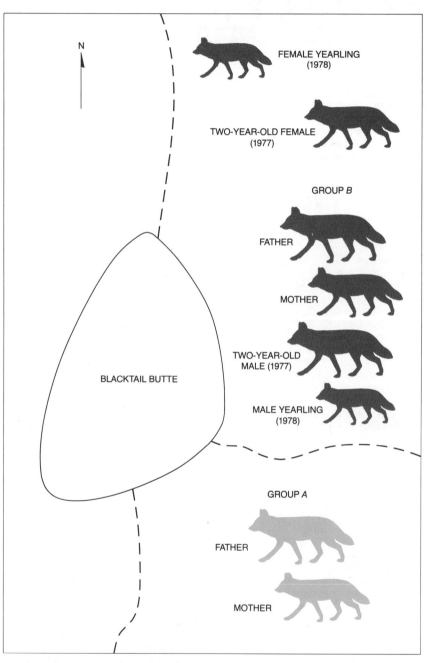

TWO COYOTE GROUPS inhabiting contiguous home ranges in the area around Blacktail Butte had access to significantly different amounts of elk carrion in the winter of 1978–79. (A home range is defined as the area an individual or a pack travels routinely in the course of its daily activities.) The home range of Group *A* held 17 percent of the available carrion, whereas the home range of Group *B* held 83 percent. As is shown here, the sizes of the groups differed accordingly: Group *A* (*color*) consisted of only a mated pair, all the young from previous years having dispersed; Group *B* (*gray*) consisted of a mated pair, a two-year-old male born to them in 1977 and a male yearling born to them in 1978. (One of the advantages of pack living may be that a breeding female receives help in caring for her young; the two-year-old in Group *B* helped to raise its siblings born in 1978.) Pack also included two hangers-on: a female born to the mated pair in 1978 and a female believed to have been born to them in 1977. These coyotes rarely interacted with their parents or siblings but were allowed to remain near them.

ing considering the wealth of ungulate carrion in it.

The sizes of coyotes' home ranges and territories vary markedly, although not consistently, with the locale, the season and the year and also with the age and the sex of the individuals. When we measured the home ranges of 10 adults in the Blacktail Butte area, we found that the average size was 21.1 square kilometers, with no discernible differences according to sex. When we classified the home-range sizes according to the coyotes' social groupings, however, we found that solitary individuals and mated pairs, which are excluded from carrion in winter, have a larger home range, with an average size of 30.1 square kilometers. Pack members, which defend a food resource in winter and tend to remain in their own territory, have an average home range of only 14.3 square kilometers. The sizes of pack members' home ranges also show considerably less variation, probably because of the clumped distribution of ungulate carrion.

Pack living confers advantages not only in the defense of food resources against competitors but also in re-productive activities. Coyotes generally mate in the period from January to April, the date varying from one locale to another. The female coyote's pregnancy rate, her productivity and her pups' rate of survival are clearly related to the general state of her health, which in turn is closely linked to the quantity and quality of the food available to her before and during pregnancy, that is, to the winter food supply. Therefore the increased ease with which pack members are often able to locate food items may represent an important reproductive advantage. Moreover, when we examined the amounts of time coyotes invest in other types of activity in winter and summer, we made an interesting discovery.

Coyotes typically are active in the early morning and early evening, but when we compared the time 50 coyotes (35 of them marked) devoted to hunting and resting, we found that in winter, when carrion is available but the food supply is usually low, much less time is spent hunting and considerably more is spent resting than is the case in summer, when small rodents are readily available but must be found, caught and killed. The higher ratio of resting time to hunting time may be generally beneficial for pregnant females, which must conserve energy for the nutritional demands placed on them during the nine-week gestation period and afterward. (There are six pups in an average coyote litter, and they are altricial, or dependent, at birth, requiring feeding and protection for the first few months of their life.) If females living in packs are able to spend more time resting than females living alone with their mate, then the pack-living females might reproduce more successfully. Moreover, as we have mentioned, females living in packs are more likely to receive help in raising their offspring.

Our findings about the pack-living adaptation of coyotes are supported by data gathered for golden jackals and hyenas, and we have been able to draw some general conclusions that should be tested with other species of carnivores. We have found that in situations where there are "haves" and "have-nots" with respect to the winter food supply (that is, individuals living in an area where a food resource is large and clumped as opposed to individuals living in one where the resource is scarce) the haves (1) are more social and cohesive than the have-nots, (2) are territorial and will defend the food resources, (3) have a more compressed home range, (4) are subject to higher rates of intrusion by members of the same species on the areas where the food is clumped and (5) in winter are able to travel less and so rest more. And the advantages of pack living can include any of the following: (1) food can be more successfully defended, particularly in winter; (2) food items can be more readily located; (3) individuals, particularly sexually mature females, can conserve energy needed for reproduction and care of the young, and (4) help, in the form of feeding and protection, can be provided for the young by individuals other than parents (most likely older siblings). Whether or not pack living confers an advantage in the acquisition of large prey remains an open question.

So far we have mainly discussed the pack-living adaptation to defendable food resources, but solitary living is also an adaptation to a particular food resource. For the coyotes we observed from Blacktail Butte the resource is rodents: prey items that coyotes cannot defend against other coyotes and that are difficult to share except with pups. Our studies have shown that even coyotes living in cohesive groups become temporarily solitary when they are hunting rodents. Hence just as it is important to study the various patterns of behavior associated with the group defense of territory and food, so it is important to study the various patterns of behavior associated with solitary predation. Not much is known about how wild coyotes

RELATIVE AMOUNTS OF TIME that coyotes in the area of Blacktail Butte devoted to the activities of hunting (*color*), traveling (*light gray*) and resting (*dark gray*) in different seasons are shown. In winter, when the coyotes depended mainly on elk carrion, the animals hunted less and rested more than they did at other times of the year. Coyotes generally mate in the winter months, and their relative inactivity in this season may be beneficial for the breeding females. A comparison of the winter activities of traveling and resting for mated pairs living in packs and those living alone reveals additional energy savings for the former (*see illustration on page 76*). Percentages are based on 668 coyote-hours of observation (one coyote-hour is defined as observation of one coyote for one hour) from September, 1977, through August, 1979.

locate and capture prey, but we have done several experiments to throw some light on this type of behavior.

To begin with, the process by which any predator locates prey is complex, and different species of carnivores go about the task quite differently. Visual, auditory and olfactory stimuli are all clearly important and in nature probably interact to elicit the predator's response to the prey. It is interesting, however, to try to determine the relative importance of these three types of stimuli for coyotes and to try to relate such findings to the natural history of the species. The experiments required for the purpose are best done with captive coyotes, under conditions in which external stimuli can be rigorously controlled.

In the first set of experiments, conducted at Colorado State University in collaboration with Philip N. Lehner, coyotes were placed in a small room 30 meters square with a hidden rabbit. The time individual coyotes needed to find the rabbit with all possible combinations of the three types of stimuli was measured. Visual stimuli were suppressed by eliminating all light from the room (in which case the coyotes were tracked by means of infrared motion-picture photography), auditory stimuli by using a dead rabbit as prey and olfac-

tory stimuli by either blowing a masking odor into the room (the odors from a rabbit colony) or by irrigating the coyote's nasal mucous membranes with a zinc sulfate solution.

The results of the experiments showed that when visual cues were present, the absence of auditory or olfactory ones led to only minor changes in the duration of the coyote's search for its prey. For example, with all three stimuli available the average search time was 4.4 seconds; with nothing but visual cues available the figure rose only to 5.6 seconds. With visual cues removed and only olfactory and auditory ones present, the average search time rose to about 36.1 seconds, or eight times the duration with all three types of stimuli. When auditory cues alone were present, the search time decreased slightly, to an average of 28.8 seconds; when olfactory cues alone were present, it went up to 81.1 seconds. With all three types of stimuli suppressed, it took the coyotes an average of 154.8 seconds, or more than 2.5 minutes, to find the prey by means of touch.

Thus under these experimental conditions the senses that facilitate the location of prey for the coyote are, in decreasing order of their importance, sight, hearing and smell. The fact that

vision is of primary importance is confirmed by the results of another series of experiments in which coyotes were presented simultaneously with a hidden rabbit making sounds (breathing, rustling and so on) and a visible rabbit making no sound. The visible rabbits were without exception captured first. The coyote probably evolved on open plains covered with low-growing grasses, where prey would be highly visible, and its heavy reliance on vision is presumably the result of adaptation to this habitat.

In order to replicate the coyote's natural hunting environment more closely a similar set of experiments was run outdoors in a large fenced-in area (6,400 square meters) at the Maxwell Ranch, owned by Colorado State University. With the larger search area and the larger number of distracting factors outdoors the average search times were all higher, but once again vision proved to be the most important sense in locating prey. Here, however, smell proved to be more important than hearing: the coyotes could find the rabbits faster with visual and olfactory cues present (when they needed an average of 34.5 seconds) than with visual and auditory cues present (when they needed an average of 43.7 seconds). Similarly, with only olfactory stimuli present the coyotes took an average of 72.7 seconds to locate the prey, and with only auditory stimuli the average search time rose to 208.8 seconds. When all three types of stimuli were present, the average search time was 30.1 seconds; when all three were suppressed, the average rose to about 22.2 minutes.

The differences between the results of the indoor experiments and those of the outdoor ones can be explained by taking into account the effects of the wind. Airborne olfactory stimuli are clearly important directional cues to a hunting coyote, as is indicated by the fact that outdoors, where smell was more important than hearing, 83.9 percent (47 out of 56) of the approaches to the rabbit were made from the downwind side. Similarly, at our study site in Grand Teton National Park we found that 74.9 percent of all the approaches we observed to mice by wild coyotes were from the downwind side. In addition, in the outdoor experiments where only olfactory cues were available to the coyotes, a significant correlation was observed between wind velocity and approach distance, or the distance at which a hunting coyote becomes aware of the location of its prey. More precisely, as the wind velocity increased the approach distance increased as well, so that when the wind was 10 kilometers per hour, the approach distance was about two meters, whereas when the wind rose to 40 kilometers per hour, the

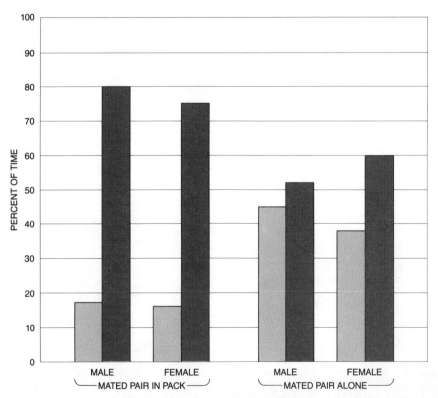

MATED FEMALE IN A PACK spends significantly more time resting (*dark gray*) and significantly less time traveling (*light gray*) in winter than a female living alone with her mate, as is shown by this chart comparing these two activities for the breeding male and female in Group A (the mated pair) and Group B (the pack) in the vicinity of Blacktail Butte (*see illustration on page 74*). Females living in packs have not been observed to reproduce more successfully than other females, but it appears that if food became a limiting factor, then the pack-living females' substantial net energy savings might give them a reproductive advantage.

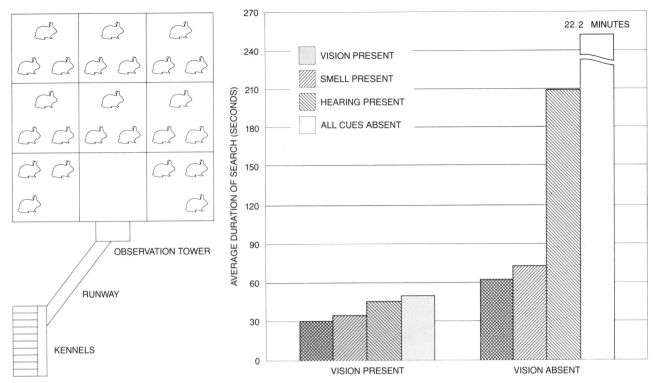

EXPERIMENTAL SETUP for determining the relative importance of the senses of vision, smell and hearing for coyotes in locating prey is shown at the left. In each trial a rabbit is placed at random at any one of 24 possible locations in a large outdoor enclosure (6,400 square meters); a coyote is admitted to the enclosure, and the time required for it to find the rabbit is recorded. The procedure was repeated for five coyotes with all possible combinations of the three types of sensory stimuli present. Visual cues were eliminated by testing coyotes on a dark night (and observing them with a "starlight scope," which intensifies available light); auditory cues were eliminated by using a dead rabbit as prey, and olfactory cues were eliminated by irrigating coyotes' nasal mucous membranes with zinc sulfate solution. Average time required to locate prey under each condition is shown at right. Results of trials with visual cues present (color) have been separated from those with visual cues suppressed, showing that in locating prey coyotes' most important sense is vision. Hearing is least important.

approach distance increased to about five meters.

Hence although the coyote seems to depend most heavily on vision when it is hunting, it appears to have effective backup systems that can be relied on when certain types of sensory cues are absent or inadequate. When prey are visible, pursuit based on visual cues is most likely to start before olfactory or auditory cues can come into play, but when the prey is well hidden, the coyote probably relies on some combination of olfactory and auditory cues. (The exact combination probably depends on the wind conditions and the amount of noise made by the prey.) Coyotes are highly efficient predators and can clearly switch back and forth between these various hunting modes in order to take maximum advantage of whatever the environmental conditions are at the time.

How does the coyote actually kill the prey it locates? Information on the subject may be useful not only to biologists interested in the comparative and evolutionary aspects of predatory behavior but also to those concerned with the control and management of predators. Here it will be most convenient to distinguish between prey animals that

are smaller than the coyote and those that are larger. (Coyotes do occasionally prey on large live animals, although as our observations of the coyotes in the area of Blacktail Butte indicate, this form of predation is rare.)

To begin with, we have observed seven distinct activities that can be included in the predatory behavior of a coyote when its prey is a small animal such as a rodent. In sequence they are long-distance searching (in which the coyote traverses large areas and scans the ground cover for a sign of prey), close searching (in which the coyote pokes around in the ground cover), orientation (in which the coyote assumes an alert posture, perhaps sniffing or pricking its ears to determine the exact location of detected prey), stalking (in which the coyote slowly and stealthily approaches its prey), pouncing (in which the coyote first rears up on its hind legs and then falls forward on its front legs to pin the prey to the ground), rushing (in which the coyote makes a rapid dash toward the prey) and finally killing. A coyote generally kills a rodent by biting it in the area of the head, and in many cases the coyote will also shake the prey vigorously from side to side.

It is important to understand that not all these activities are always included in

a single predatory sequence. For example, we found that if the prey is a smaller rodent such as a field mouse, a coyote does not usually rush the rodent but simply stalks it and then pounces on it, pinning it to the ground so that a killing bite can be delivered. When the prey is a larger rodent such as a Uinta ground squirrel or a Richardson's ground squirrel, however, the coyotes we observed rushed it in more than 90 percent of the cases and pounced only rarely.

The success of the coyote's predatory sequences in catching and killing rodents varies considerably. Our data indicate that coyotes are successful between 10 and 50 percent of the time. We have not yet identified all the variables that influence the rate of success, but ground squirrels seem to be easier to catch than mice. The hunger level of a coyote may also be important. Observations in captivity reveal that satiated coyotes often play with a rodent before killing and eating it, and frequently the rodent escapes. Similar observations have been made in the field.

We also wondered whether the predatory skills of coyotes improve with age, and so we compared the time that nine young coyotes from three to six months old and 15 adults spent in the activities of searching, orienting and stalking when they were hunting mice or ground squirrels. The adults, it turned out, spent less time searching and orienting, and in addition the times adults devoted to these activities were much less variable than those of the pups. There was no difference in the time spent stalking, however, an activity to which coyotes in both age groups devoted an average of about 5.5 seconds. Therefore it would appear that the pups are less effective than the adults in locating their prey, but once the prey has been located coyotes in either age group will stalk briefly and then go in for the kill. Studies of coyotes in captivity also reveal that pups only 30 days old are capable of carrying out a successful predatory sequence on a mouse. In other words, although coyotes of that age rarely have an opportunity to kill a small rodent in the wild, they clearly have the ability.

Turning to the subject of how coyotes kill large wild prey, such as sheep, deer, elk and moose, there are for a number of reasons few observations from which useful generalizations can be drawn. To begin with, coyote kills are often indistinguishable from those of other wild predators or even domestic dogs. Moreover, it has been noted that most healthy ungulates living in the same locale as coyotes are able to defend themselves against a single coyote, so that instances of such predation are rare and hence difficult to observe. The few data that do exist indicate that two or more coyotes are usually required to take down, say, a healthy adult deer. In most cases coyotes appear to kill either young ungulates or weak ones, typically by attacking the head, neck, belly and rump. It is generally believed coyotes do not have any significant detrimental effect on wild ungulate populations.

The effects of coyote predation on domestic sheep are less clear-cut, which brings us to a more controversial aspect of coyote biology, namely the management and control of coyotes. Coyotes are said to have a significant detrimental effect on the sheep industry, and as a result for a century coyotes have been a particular target of predator-control programs. At present large amounts of time, energy and money (in many cases from public funds) are being devoted to such efforts. The returns on the investment are small, in terms both of reducing coyote populations and of preventing livestock losses and damage. The failure of the control and management programs is due essentially to the lack of sufficient background information on the behavioral and population dynamics of coyotes.

Indeed, very little is known about the predatory habits of wild coyotes with regard to domestic sheep. Guy Connolly and his colleagues at the United States Fish and Wildlife Service have found that even when coyotes are confined with sheep, their predatory behavior is surprisingly inefficient. In these experiments coyotes killed sheep in only 20 out of 38 encounters. Moreover, both the average time that elapsed before the coyotes attacked the sheep (47 minutes) and the average time that elapsed before the sheep were killed (13 minutes) were quite long, totaling an hour. The defensive behavior of the sheep deterred the coyotes in only 31.6 percent of the cases, and so it is understandable that the coyotes would take their time before killing the sheep. Of course, there are no instances of such inefficient predation in natural predator-prey interactions, where the prey either flees or actively fights off the predator as long as it can. It is clear, however, that sheep, which have been subjected to artificial selection by the great domesticator *Homo sapiens,* have been left virtually defenseless against predation.

Coyotes do kill sheep, then, as well as other livestock and poultry. Many studies have shown, however, that factors other than coyote predation can cause considerably heavier losses. For example, it was reported in a recent study that in the early 1970's the value of the losses of ewes and lambs in the state of Idaho amounted to $2,343,438. Of this total 36 percent could be attributed to disease, 30 percent to unspecified causes and 34 percent to predation; only 14.3 percent of the losses could be attributed to predation by coyotes. Moreover, there are data to indicate that not all coyotes are sheep killers and that the indiscriminate killing of coyotes in areas where sheep are being killed is an ineffective method of control. A recent study of livestock predation in 15 Western states issued by the Animal Damage Control Program of the Department of the Interior concluded that the relation between such predation and the population dynamics of coyotes is obscure.

In a sense the coyote is victimized by success: it is threatened because it takes advantage of livestock that have been robbed of most of their defenses by shortsighted practices of domestication. It is to be hoped that in the future defensive behavior will be bred back into livestock. For the present one can only assume that the failure of predation control is due to a lack of basic knowledge about predatory species, a problem that can be remedied by further studies of behavior and ecology of the kind we have described here.

We have found the coyote to be a particularly good subject for such investigations. Further field study will be needed to determine to what extent our findings can be applied to other coyote populations, to closely related species and to carnivores in general. In the meantime coyotes should be appreciated as animals that have adapted remarkably well to the pressures exerted by their environment, including harassment by man and the severe restriction of their natural habitats.

Dolphins

by Bernd Würsig
March 1979

*These descendants of land mammals that took to sea
have a large brain, learn quickly and exhibit a rich
vocal repertory. Yet lack of evidence leaves open the
question of how intelligent they are.*

Because dolphins have big brains, are quick to learn tricks devised by human trainers and exhibit a rich repertory of vocal signals, they are widely reputed to have a level of intelligence unmatched by any other animal and perhaps even the equal of man's. On the basis of observations of dolphins in several of their natural habitats I believe the effort to position them firmly in the spectrum of animal intelligence is premature. In the present state of marine technology it is simply impossible for human observers to spend more than brief and isolated moments with an animal that lives in the ocean and moves rapidly over great distances. When a great deal more is known about the behavior of dolphins, the question of their intelligence will answer itself. At present the most one can say is that dolphins are gregarious herding animals, comparable in their individual and social behavior to more easily observed herding and flocking mammals on land.

Dolphins evolved at least 50 million years ago from land mammals that may have resembled the even-toed ungulates of today such as cattle, pigs and buffaloes. After taking to the sea dolphins became progressively better adapted to life in the water: their ancestral fur was replaced by a thick coat of blubbery fat, they became sleek and streamlined, they lost all but internal remnants of their hind limbs and grew a powerful tail, their forelimbs were modified into steering paddles and apparently as a further aid in steering and stabilization many species evolved a dorsal fin.

More than 30 species of dolphins can be identified. They belong to the suborder Odontoceti, or toothed whales, of the order Cetacea. (The quite similar porpoise, which is often confused with the dolphin, is also a member of the whale family; it is distinguished from the dolphin mainly by having a less beaklike snout and also by its laterally compressed, spadelike teeth. It should be noted, however, that many American students of marine mammals refer to all small odontocete cetaceans as porpoises, regardless of their physical charac-

teristics.) A few dolphin species live in fresh water, but most species have an ocean habitat. The freshwater species travel in small groups or are nearly solitary, whereas the ocean species (such as the Pacific spotted dolphin) may congregate in aggregations of several thousand. Such numbers are reminiscent of buffalo herds in North America and grazing animals on the Serengeti plains of Africa; one wonders about behavioral and ecological similarities between the dolphins and their distant terrestrial relatives.

In general highly social mammals have complex social signals and a rich behavioral repertory. Hence they can interact with other members of their group in sophisticated ways. Examples include signals of aggression that are useful in establishing and maintaining dominance hierarchies, signals for courtship, warning sounds or movements at the approach of a potential attacker and many other signals that contribute to the functioning of the group and its individual members.

The effort to accumulate data on the behavior and social systems of dolphins is made difficult by the fact that most of their communication goes on below the waves. It is extremely difficult to approach a group of dolphins in a boat and to stay with them long enough to begin to understand their social system. All one sees is a group of dorsal fins as the animals surface to breathe, and then they are lost to view as they move on underwater. Even in the rare circumstances when the water is calm and clear and the dolphins can be seen for more than a few minutes, the proximity of a boat may disturb them, so that it becomes difficult to separate what is natural in their behavior from what is unnatural, being merely a reaction to the boat.

As a result of these problems and others most of the early observations involved dolphins in captivity. Although captivity must be an awkward situation for animals that are accustomed to a life with few physical boundaries, the captive dolphins nonetheless yielded useful

data simply because observers could watch them for extended periods of time. The best-known work of this kind was done by Margaret C. Tavolga of Marineland of Florida. She observed a group of 12 bottlenose dolphins (*Tursiops truncatus*) in a large tank for a total period of about five years.

Tavolga found that the group had a definite dominance hierarchy. The one adult male, which was the largest animal in the group, was more aggressive and less fearful than any of the females, subadult males or young dolphins. In general the larger animals were dominant over the smaller ones.

Similar data were obtained by Gregory Bateson at the Oceanic Institute of Hawaii. He found a dominance hierarchy in a group consisting of two spotted dolphins (*Stenella attenuata*) and five spinner dolphins (*S. longirostris*). The largest male threatened other dolphins (by lunging at them or showing his teeth) but was never threatened himself. The second-ranking dolphin, also a male, threatened the animals below him, and so on down the line. Bateson's findings showed, as the findings of other workers have shown, that the hierarchy is not as strict as it is among some other mammals. For example, the lowest-ranking male was still able to mate with a female without being challenged by the largest male.

If the dominance hierarchy is not necessarily related to the access of the males to the females (as it is, for example, in the harem of the elephant seal), one wonders what its function is in wild dolphin populations. Kenneth S. Norris and Thomas P. Dohl of the University of California at Santa Cruz have speculated that the function may be to organize the members of the group to deal with a variety of situations. For example, threats and chases by the larger dolphins could cause the smaller females and young animals to be herded into the center of the group, where they would be better protected from such potential predators as sharks and killer whales. Norris believes he has seen such struc-

THREE STAGES OF A LEAP are demonstrated (unintentionally) in the Sea of Cortés (the Gulf of California) off the coast of Mexico by three dolphins of the species *Delphinus delphis.* The repertory of leaps, spins and somersaults executed by dolphins is richly varied.

HERD OF DOLPHINS of the species *Delphinus delphis* was photographed leaping in the Atlantic. Dolphins are social mammals that sometimes congregate in quite large herds, as they are doing here, but more often they are found in subgroups of perhaps 20. Dolphins are also air-breathing animals, so that their leaps serve in part to enable them to breathe. The animals also leap during play and hunting.

turing in spinner dolphins and spotted dolphins.

For such a system to evolve it is helpful and perhaps necessary that the animals possessing it be genetically related. W. D. Hamilton of the Imperial College of Science and Technology and other biologists have argued that closely related animals should tend to protect one another more than they protect distant relatives, since close relatives share more genes. If a mother saves her offspring by herding it inside the group while putting herself in at least some danger, her apparent altruism will be adaptive for the group because a significant proportion of her genes will be preserved.

If dolphins herd group members in this way, it is likely that at least some of the animals in a group are related. Dolphins have also been seen to help an ailing member of the group reach the surface to breathe and to protect a group member from predators or other dangers. These patterns of behavior have often been cited as evidence of humanlike altruism and of great intelligence. It appears more likely that they represent an outgrowth of an evolved tendency to help related individuals.

Unfortunately the degrees of consanguinity in a wild dolphin population are not known. Only recently has any idea at all of social structure among dolphins been gained, and the knowledge is complicated by the existence of many different species and many more separate populations within species. From terrestrial mammals it has been learned that the social system represents in part an adaptation to the population's habitat. For example, wolves that live mainly on deer tend to travel in small packs, whereas wolves that hunt moose are found in larger and more highly organized societies. This difference is apparently related to the need for a coordinated effort by several animals to successfully bring down the larger prey.

A few excellent studies that make comparisons possible among dolphin species and populations have been done recently. Norris and Dohl studied Hawaiian spinner dolphins from sea cliffs and from underwater. They found that these highly social mammals traveled in schools averaging 25 members. The structure of the school varied during the day in a predictable manner. In the morning the dolphins moved slowly and in tight groups, with individuals almost touching. They appeared to be resting. Later they became increasingly active, swimming faster, with individuals leaping clear of the water in the spins, somersaults and other displays for which dolphins in oceanariums are famous.

At such times the schools became more spread out, with animals often as much as 20 meters apart. Moreover, groups tended to join, so that 50 or more members might constitute the expanded school, with all the animals moving in the same direction. As night approached, the school moved several kilometers away from the shore, entering deeper water and beginning deep dives in order to feed on fish several hundred meters below the surface. The array consisted of many widely spaced groups within an area with a diameter of several kilometers. Because Norris and Dohl were able to recognize some individuals, they found that a given small group of individuals tended to stay together but often shifted as a unit from one school to another.

Norris and Dohl suggested that dolphins may form close groups while they are resting so that they can employ the combined sensory abilities of all the individuals in the school to scan the environment and to detect potential danger. It is well known that dolphins can scan the water by echo location over much greater distances than would be possible by eyesight. (In echo location a dolphin projects high-frequency sounds in short pulses, much as bats do. The sounds bounce off objects, and the echoes give back information on the distance, size, shape and even texture of the object.) Norris and Dohl's hypothesis about combined sensory abilities, particularly during rest, relies in part on the ability of dolphins to get information from echo location. Presumably each dolphin

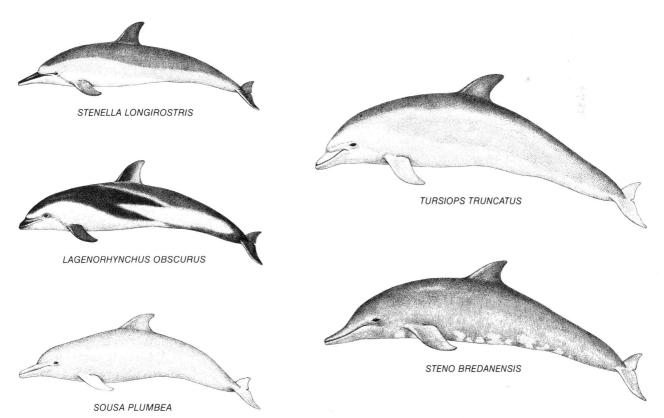

STENELLA LONGIROSTRIS

LAGENORHYNCHUS OBSCURUS

SOUSA PLUMBEA

TURSIOPS TRUNCATUS

STENO BREDANENSIS

FIVE DOLPHIN SPECIES are portrayed to indicate their differences in shape and marking. They are the bottlenose dolphin (*Tursiops truncatus*), the spinner dolphin (*Stenella longirostris*), the South Atlantic dusky dolphin (*Lagenorhynchus obscurus*), the Indo-Pacific humpback dolphin (*Sousa plumbea*) and the rough-toothed dolphin (*Steno bredanensis*). All of them are drawn to the same scale. About 25 other species of dolphins are known. Porpoises, which are quite similar, have a blunter snout and are ordinarily shorter and fatter.

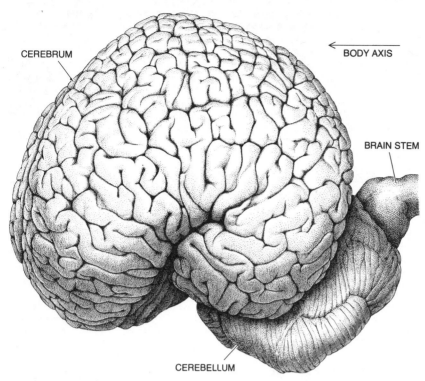

CEREBRUM

BODY AXIS

BRAIN STEM

CEREBELLUM

BRAIN OF DOLPHIN is depicted for the species *Tursiops truncatus,* the bottlenose dolphin. A dolphin brain weighs about 3.5 pounds (1.59 kilograms); a typical human brain weighs about three pounds (1.36 kilograms) but is larger than the dolphin's in proportion to body weight.

also represent a form of communication. A leap is usually followed by a loud slap or splash as the dolphin enters the water. Such sounds travel fairly long distances underwater and may signify the presence of the leaper to others. Indeed, groups of some dolphin species sometimes converge on an active, leaping school from a distance of several kilometers. Nighttime feeding in deep water is also attended by much leaping and loud splashing. At this time the members of a group are quite widely separated, and the assumption once again is that leaping may serve to communicate location and possibly information such as the number of other dolphins nearby and what they are doing.

Graham Saayman and C. K. Tayler of the Port Elizabeth Museum in South Africa studied Indian Ocean bottlenose dolphins (*Tursiops aduncus*), and my wife Melany Würsig and I studied South Atlantic dusky dolphins (*Lagenorhynchus obscurus*). Both species have habitats similar to the habitat of the Hawaiian spinner dolphin. The habitats are coastal-pelagic, meaning that all three populations can often be seen and studied from the shore but that they also move far from the shore, usually to feed. All three populations exhibited similar patterns of behavior and movement. It therefore seems reasonable to say that the habitat of these marine mammals is largely responsible for their way of life.

in a closely organized school can hear the echo-location sounds made by other members of the group. Therefore even though any given individual might not make many sounds, much information about the environment would be rapidly and efficiently disseminated to all. It is also probable that a resting group swims

close to shore in order to be in shallow water that is not frequented by large deepwater sharks.

During periods of alertness the spacing of the spinner dolphins increases and the animals do a great deal of leaping. This activity may in part be play, as many people have suggested, but it may

This assertion can be examined by studying dolphins in a different environment. Three such environments can be found: deep-ocean, coastal and freshwater. The most thoroughly examined populations have been coastal ones. Susan H. Shane of Texas A & M University observed Atlantic bottlenose dolphins off Texas; A. Blair Irvine, Randall S. Wells and Michael Scott of the University of Florida observed them off Florida; Melany Würsig and I observed them off Argentina, and Saayman and Tayler observed Indo-Pacific humpback dolphins (of the genus *Sousa*) off South Africa. Again major similarities were evident. They can be summarized by a description of our study of bottlenose dolphins off Argentina.

The bottlenose dolphin is a coastal species in many parts of its worldwide range. Hence it can be observed readily from the shore. For a period of 21 months my wife and I observed a school of bottlenose dolphins that passed close to the shore (always within a kilometer) in water less than 40 meters deep. We studied them by observations from coastal cliffs and from a small rubber boat, by underwater recordings of their sounds, by photographing their dorsal fins in order to recognize individuals and by tracking their movements with a surveyor's transit on the shore. (With a transit one can determine precisely the

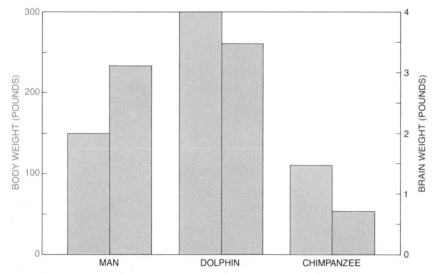

BRAIN WEIGHT AND BODY WEIGHT are compared for man, the dolphin and the chimpanzee. Man's brain weighs about 2 percent of his body weight, the chimpanzee's less than 1 percent and the dolphin's slightly more than 1 percent. Although rough ideas of comparable intelligence can often be obtained by comparing the ratio of brain weight to body weight, other factors such as the length of the body, the manner in which limbs are used and the complexity of the brain must also be considered. It is not possible to say on the basis of a simple comparison of brain weight to body weight that one kind of organism is more intelligent than another.

location, movement and speed of wild animals. The method was developed by Roger Payne of the New York Zoological Society and has proved to be useful in studies of the coastal movements of marine mammals.)

The school ranged in size from eight to 22 dolphins; the average was 15. Some individuals stayed together consistently; others were at times absent. Thus the population was made up of subunits of shifting membership.

Eventually our sightings were all of animals we had identified previously. We concluded that we knew all or nearly all 60 or so dolphins that at one time or another were part of the group. Although all the animals in the larger group interacted to some degree, we never saw the entire group together at one time. It is therefore likely that other subgroups, also composed of individuals known to us, traveled near other shores. Apparently they cover a large range; we once sighted a subgroup of known animals more than 300 kilometers south of our base. It is probable that the size of a subgroup represents an optimum balance among the number of animals needed for defense against predators, the number needed for efficient feeding and the number best suited for social interaction, reproduction and the survival of young.

In the fall, winter and spring the bottlenose dolphins seemed to feed on schools of anchovies at about midday in water from 15 to 35 meters deep. At such times they advanced as a spread-out school, each animal being separated from the next one by as much as 25 meters. After several minutes in this formation they began to dive and mill around in one area. From comparison with the behavior of dusky dolphins, which in this region feed almost exclusively on anchovies, we deduced that the bottlenose dolphins were herding schooling anchovies to the ocean surface and feeding on them there. The spread formation before feeding probably serves the purpose of acoustically scanning as large an area as possible in the search for anchovies.

This cooperation among members of a school is quite different from what the dolphins do in the summer. Anchovies are not present in the waters off Argentina during the summer, and so the dolphins feed mainly on large solitary fishes living among rocks near the shore. At such times the dolphins move in water that is from two to six meters deep, and they spread out in a line that is longer than it is wide, with every animal at essentially the same depth and as close to the shore as possible. In this formation individual dolphins nose their way among the rocks and poke into crevices as they search for their prey.

Although the average size of subgroups was 15 dolphins, the average was lower (14) during the summer than it was during the winter (20). It can be argued that fewer dolphins are needed for successful individual feeding near the shore. Indeed, it is possible that the resource—large fishes inhabiting crevices—is limited and that a group of about 14 dolphins in one area at one time represents the upper limit of the area's carrying capacity. The limit is higher when dolphins are feeding cooperatively on schooling fish.

Other reasons for the seasonal fluctuation in the size of dolphin subgroups must also be considered. For example, one could argue that the size of subgroups increases because dolphins are more susceptible to killer-whale attacks when they feed farther from shore and that with more animals per subgroup they can protect themselves better. Perhaps the fluctuations reflect seasonal peaks in mating and calving. On the basis of the information now available, however, food seems to be an important determinant of the size (and presumably the composition) of subgroups of bottlenose dolphins.

What can be said about the internal structure of subgroups? The fact that individual dolphins shift about from one subgroup to another suggests that the animals have what is known among mammalogists as an open society. Among terrestrial mammals the African chimpanzee (*Pan troglodytes*) exhibits a quite similar system. Certain African ungulates (hoofed mammals) also have a degree of openness, with individuals moving frequently from subgroup to subgroup within a more rigidly defined herd. Among chimpanzees the variations appear to be in response to the availability of food; the animals search for food in small units but aggregate into larger units when the food has been found. Melany Würsig and I have found a similar situation among dusky dolphins, but it is not yet clear to what extent a patchy distribution of food might govern variations in the size of schools of bottlenose dolphins.

Notwithstanding the tendency to openness, groups of five or six dolphins were almost always observed traveling together. One such group included a notably large animal and a second adult that traveled with a calf during the entire 21-month study. I therefore believe the second adult was a female. The large

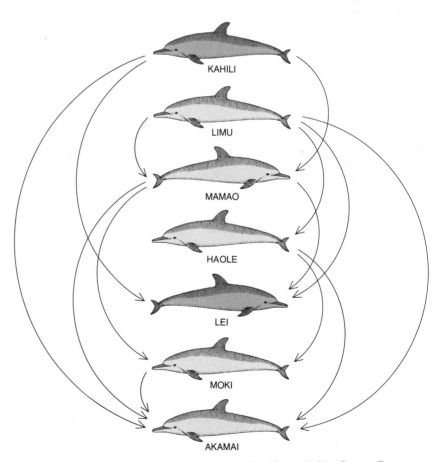

DOMINANCE HIERARCHY in a group of captive dolphins studied by Gregory Bateson at the Oceanic Institute of Hawaii is depicted. The group consisted of two spotted dolphins (*dark gray*) and five spinner dolphins (*light gray*). In this drawing males face to the left, females to the right. The arrows indicate threats (threatening entails showing teeth or lunging) made by one animal to another. The largest male, Kahili, and the male Limu were never threatened. Although Akamai, the smaller male dolphin at the bottom of the hierarchy, never threatened any other member of the group, he copulated with the female Lei on at least one occasion.

animal was the only one that consistently slapped its tail against the surface of the water when a boat approached. Experienced dolphin trainers say that tail slapping can signify anger, aggression or warning. It is tempting to say that the large dolphin was in some sense the leader of the group, but I do not have enough information about the social system and behavior of dolphins to substantiate the hypothesis.

Two other groups of five and six dolphins that we always found together consisted of adults of approximately the same size. It is possible (again I conjecture) that the groups were made up of nonbreeding members of the population, much as bachelor herds of elephants travel together.

The most detailed findings on the age and sex composition of dolphin schools have come from the work of Ir-

vine, Wells and Scott on Florida bottlenose dolphins. They captured 47 dolphins and put tags and other identifying marks on them in order to recognize them later in their natural habitat. As they tagged the animals they were able to determine the sex and size of the individuals.

Once the dolphins had been released Irvine, Wells and Scott found that the home range of the resident herd covered about 85 square kilometers. Females and calves often traveled in groups that included only a few adult males or none. Such males tended to associate more often with calfless females than with mothers and young, and they rarely associated with subadult males. The subadult males were at times found in bachelor groups far from the other dolphins. Several females were seen with their calves for as long as 15 months. Hence a

strong social bond exists between mother and calf, probably continuing long after weaning. No such long-term association between a male and a calf has been observed.

It should be emphasized that the social relations of dolphins are not clear-cut or immutable. They are highly variable. Nevertheless, a few major features are apparent. The bonds between mother and calf are strong; the bonds between male and female and male and calf appear to be less so. This comparison suggests that mating is somewhat promiscuous. Subadult males may be excluded from the normal social routine but subadult females are not, which suggests that adult males may copulate more with various females than females do with different males. Such a relation is indicative of a polygynous mating system, which is also common among terrestrial mammals.

Less can be said about the behavior of dolphins than about their social relations, even though a considerable body of accounts of behavior has been built up from observations of dolphins in nature and in captivity. I believe not enough is yet known to support any firm and broad statements. Still, a few major examples of behavior can be cited. Bottlenose dolphins (and other species) appear to engage in courtship and copulation throughout the year, as is often indicated in the wild by belly-to-belly swimming. Yet bottlenose dolphins and some other species have a definite yearly calving peak (sometimes two peaks). Among the dolphins we observed off Argentina all the calves were born in the summer. This finding indicates that a physiological change in the male or the female causes conception to occur in a limited period. Such a change has been documented in seasonal increases in the weight of the testes in the males of several dolphin species.

Yearlong mating also implies that courtship may have more than a sexual connotation. Several investigators have suggested that such interactions may also serve to define and strengthen social hierarchies and bonds. The argument is reinforced by the frequent homosexual activity seen at least among captive dolphins. Future studies may show that "homosocial" might be a better term. A carry-over of sexual signals to dominance hierarchies is seen in many other mammalian groups.

A second behavior found in almost all dolphin species is leaping. I have mentioned that it tends to occur most often when animals are widely separated and so may have a communicative function. Bottlenose dolphins off Argentina leaped far less than dusky dolphins in the same vicinity did, even when both species were hunting fish in essentially the same manner. The bottlenose dolphins, however, moved in one school

MEANS OF IDENTIFICATION of individual dolphins is provided by the pattern of nicks and scars on the trailing edge of the dorsal fin. These photographs represent a sampling of 12 bottlenose dolphins from a group of about 50. All individuals in the group could be identified.

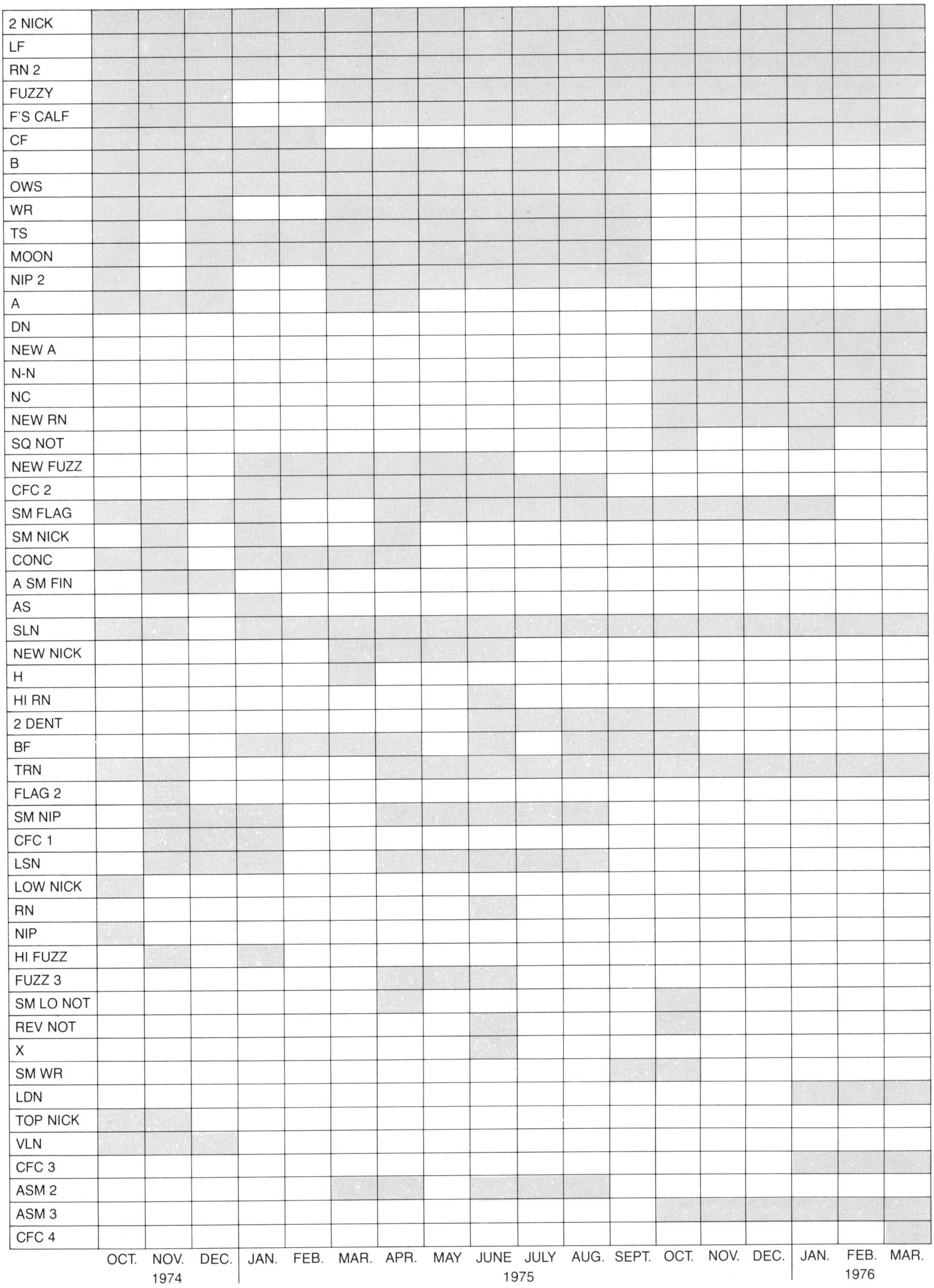

2 NICK
LF
RN 2
FUZZY
F'S CALF
CF
B
OWS
WR
TS
MOON
NIP 2
A
DN
NEW A
N-N
NC
NEW RN
SQ NOT
NEW FUZZ
CFC 2
SM FLAG
SM NICK
CONC
A SM FIN
AS
SLN
NEW NICK
H
HI RN
2 DENT
BF
TRN
FLAG 2
SM NIP
CFC 1
LSN
LOW NICK
RN
NIP
HI FUZZ
FUZZ 3
SM LO NOT
REV NOT
X
SM WR
LDN
TOP NICK
VLN
CFC 3
ASM 2
ASM 3
CFC 4
OCT. NOV. DEC.
1974
JAN. FEB. MAR. APR. MAY JUNE JULY AUG. SEPT. OCT. NOV. DEC.
1975
JAN. FEB. MAR.
1976

GROUP COHESIVENESS of dolphins is indicated by this chart of the presence and absence of 53 known bottlenose dolphins off the coast of Argentina over a period of 18 months. The designations at the left are abbreviated names the observers gave the dolphins. A bar opposite a designation indicates that the individual was seen near shore at least once during the corresponding month; a blank space means the animal was not seen during that month. A dolphin shown as having been seen may have been seen more than once in the month.

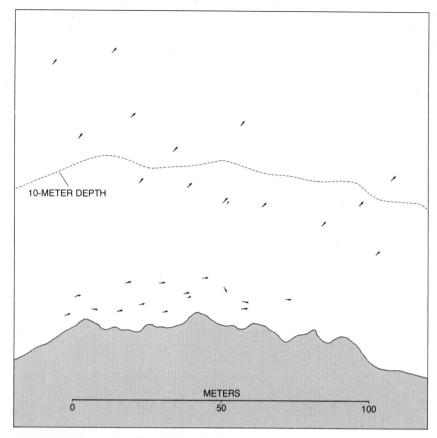

GROUP FORMATIONS of bottlenose dolphins off the coast of Argentina varied according to whether the animals were in shallow water close to shore or in deeper water farther out. In shallow water the dolphins were individually hunting rock-dwelling fishes, whereas in deep water they functioned as a group to find schools of anchovies and to herd them to the surface.

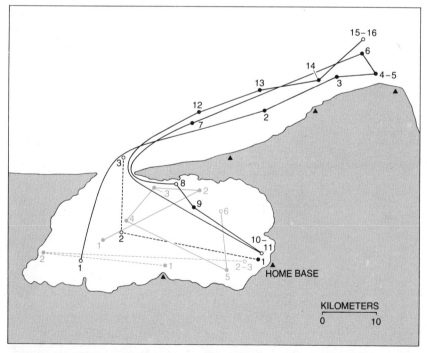

RADIO TRACKING of four dolphins off Argentina (in the Gulf of San José and the adjoining open ocean) produced these patterns of movement. Each dolphin is represented by a different track, and the number beside each circle shows the animal's position that number of days after it was fitted with a radio. Solid circles represent known positions, obtained either by triangulation from the shore points indicated by triangles or by approaching the dolphin in a boat, and open circles represent estimated positions. The animals' daily locations are given as of midday.

and so would have little cause to communicate with other members of the school by leaping. The dusky dolphins moved in as many as 30 small schools in one vicinity, and we often saw schools coalesce when leaping began. Leaps undoubtedly have other functions, such as helping to herd or catch prey, but the function of communication among members of a particular species may be important.

On the basis of the rather modest body of knowledge about dolphins, what can be said of their intelligence? Dolphins are certainly adept at learning complex tasks, as they demonstrate in their tricks in oceanariums, and they remember the tasks for years. They also have been shown to be capable of relatively abstract thinking. For example, Karen Pryor, working at the Oceanic Institute in Hawaii, trained rough-toothed dolphins (*Steno bredanensis*) to perform a new trick for a reward of fish. After several days of training they exhibited ever-different types of leaps and contortions, apparently "realizing" that the forms of behavior they had displayed previously would not be rewarded. Still, various trainers have pointed out that the same thing happens among dogs and other mammals and possibly even among pigeons, which implies that one does not have to invoke superintelligence to explain what the dolphins are doing.

Edward O. Wilson of Harvard University has suggested that the brain of the dolphin may be larger in relation to the size and weight of the body than the brains of most other mammals because of the same reputed imitative abilities that have made dolphins such a favorite with animal trainers and the public. The question then is: Why should an animal be such a superb imitator? R. J. Andrew of the University of Sussex has noted that vocal mimicry may be important to animals that often travel out of sight of one another. Individuals of a widely spaced group could then recognize other members because of an elaborate convergence of signals among the animals of one group or herd. This system, in the form of dialects, has been shown to operate among some primates and birds. It is plausible that the system evolved to an even greater degree among dolphins, which rely heavily on sound.

What about mimicry of movement? Wilson suggests that individual dolphins may imitate the members of the group that are most successful at catching fish and avoiding predation. Furthermore, it is advantageous for animals in social societies that cooperate to hunt food (as has been shown in at least some dolphin species) to know one another's movements well and for individuals to be able to take several roles in the herding of a school of fish. Wilson argues that imitation alone is enough to explain the size of the dolphin's brain and that the social signals of dolphins are probably no more sophisticated than those of most mammals and birds. In my opinion not enough is known about the social signals of dolphins to provide a basis for such a statement. Norris views the imitative powers of dolphins as not being necessarily better than those of many other mammals with brains that are smaller and physically less complex. It seems futile at present to compare the intelligence of dolphins with that of other mammals simply because of the lack of appropriate information about dolphins and the great differences between their environments and those of terrestrial mammals.

The need for better information about dolphins turns one's mind to better means of obtaining it. One possibility is to try to habituate dolphins to observers to such an extent that they will go about their daily activities as if the observers were not present. George Schaller studied mountain gorillas in this way, and Jane Goodall similarly opened a new era of work on chimpanzees. They moved with the animals and sat patiently until the animals either accepted them or simply ignored them.

How might one follow a group of dolphins in the ocean? Perhaps it is not necessary. Jody Solow of the University of California at Santa Cruz recently learned to make a sound underwater that at times called individuals of a group of nearby Hawaiian spinner dolphins to her. Her achievement opens the possibility that an investigator could eventually recognize all the members of a group, learn their social patterns and interactions and gain a better idea of their natural behavior.

VARIETY OF LEAPS performed by dolphins is suggested by these photographs of three dusky dolphins (*Lagenorhynchus obscurus*) in the open ocean. Dolphins in captivity have been observed to increase the variation in their leaps when they are receiving rewards of food.

The Social Life of Baboons

10

June 1961

*A study of "troops" of baboons in their natural
environment in East Africa has revealed patterns of
interdependence that may shed light on the evolution
of the human species.*

The behavior of monkeys and apes has always held great fascination for men. In recent years plain curiosity about their behavior has been reinforced by the desire to understand human behavior. Anthropologists have come to understand that the evolution of man's behavior, particularly his social behavior, has played an integral role in his biological evolution. In the attempt to reconstruct the life of man as it was shaped through the ages, many studies of primate behavior are now under way in the laboratory and in the field. As the contrasts and similarities between the behavior of primates and man—especially preagricultural, primitive man—become clearer, they should give useful insights into the kind of social behavior that characterized the ancestors of man a million years ago.

With these objectives in mind we decided to undertake a study of the baboon. We chose this animal because it is a ground-living primate and as such is confronted with the same kind of problem that faced our ancestors when they left the trees. Our observations of some 30 troops of baboons, ranging in average membership from 40 to 80 individuals, in their natural setting in Africa show that the social behavior of the baboon is one of the species' principal adaptations for survival. Most of a baboon's life is spent within a few feet of other baboons. The troop affords protection from predators and an intimate group knowledge of the territory it occupies. Viewed from the inside, the troop is composed not of neutral creatures but of strongly emotional, highly motivated members. Our data offer little support for the theory that sexuality provides the primary bond of the primate troop. It is the intensely social nature of the baboon, expressed in a diversity of inter-

individual relationships, that keeps the troop together. This conclusion calls for further observation and experimental investigation of the different social bonds. It is clear, however, that these bonds are essential to compact group living and that for a baboon life in the troop is the only way of life that is feasible.

Many game reserves in Africa support baboon populations but not all were suited to our purpose. We had to be able to locate and recognize particular troops and their individual members

GROOMING to remove dirt and parasites from the hair is a major social activity among baboons. Here one adult female grooms another while the second suckles a year-old infant.

and to follow them in their peregrinations day after day. In some reserves the brush is so thick that such systematic observation is impossible. A small park near Nairobi, in Kenya, offered most of the conditions we needed. Here 12 troops of baboons, consisting of more than 450 members, ranged the open savanna. The animals were quite tame; they clambered onto our car and even allowed us to walk beside them. In only 10 months of study, one of us (DeVore) was able to recognize most of the members of four troops and to become moderately familiar with many more. The Nairobi park, however, is small and so close to the city that the pattern of baboon life is somewhat altered. To carry on our work in an area less disturbed by humans and large enough to contain elephants, rhinoceroses, buffaloes and other ungulates as well as larger and less tame troops of baboons, we went to the Amboseli game reserve and spent two months camped at the foot of Mount Kilimanjaro. In the small part of Am-

boseli that we studied intensively there were 15 troops with a total of 1,200 members, the troops ranging in size from 13 to 185 members. The fact that the average size of the troops in Amboseli (80) is twice that of the troops in Nairobi shows the need to study the animals in several localities before generalizing.

A baboon troop may range an area of three to six square miles but it utilizes only parts of its range intensively. When water and food are widely distributed, troops rarely come within sight of each other. The ranges of neighboring troops overlap nonetheless, often extensively. This could be seen best in Amboseli at the end of the dry season. Water was concentrated in certain areas, and several troops often came to the same water hole, both to drink and to eat the lush vegetation near the water. We spent many days near these water holes, watching the baboons and the numerous other animals that came there. On one occasion we counted more

than 400 baboons around a single water hole at one time. To the casual observer they would have appeared to be one troop, but actually three large troops were feeding side by side. The troops came and went without mixing, even though members of different troops sat or foraged within a few feet of each other. Once we saw a juvenile baboon cross over to the next troop, play briefly and return to his own troop. But such behavior is rare, even in troops that come together at the same water hole day after day. At the water hole we saw no fighting between troops, but small troops slowly gave way before large ones. Troops that did not see each other frequently showed great interest in each other.

When one first sees a troop of baboons, it appears to have little order, but this is a superficial impression. The basic structure of the troop is most apparent when a large troop moves away from the safety of trees and out onto open plains. As the troop moves the less dominant

MARCHING baboon troop has a definite structure, with females and their young protected by dominant males in the center of the formation. This group in the Amboseli reserve in Kenya includes a female (*left*), followed by two males and a female with juvenile.

adult males and perhaps a large juvenile or two occupy the van. Females and more of the older juveniles follow, and in the center of the troop are the females with infants, the young juveniles and the most dominant males. The back of the troop is a mirror image of its front, with less dominant males at the rear. Thus, without any fixed or formal order, the arrangement of the troop is such that the females and young are protected at the center. No matter from what direction a predator approaches the troop, it must first encounter the adult males.

When a predator is sighted, the adult males play an even more active role in defense of the troop. One day we saw two dogs run barking at a troop. The females and juveniles hurried, but the males continued to walk slowly. In a moment an irregular group of some 20 adult males was interposed between the dogs and the rest of the troop. When a male turned on the dogs, they ran off. We saw baboons close to hyenas, cheetahs and jackals, and usually the baboons seemed unconcerned—the other animals kept their distance. Lions were the only animals we saw putting a troop of baboons to flight. Twice we saw lions near baboons, whereupon the baboons climbed trees. From the safety of the trees the baboons barked and threatened the lions, but they offered no resistance to them on the ground.

With nonpredators the baboons' relations are largely neutral. It is common to see baboons walking among topi, eland, sable and roan antelopes, gazelles, zebras, hartebeests, gnus, giraffes and buffaloes, depending on which ungulates are common locally. When elephants or rhinoceroses walk through an area where the baboons are feeding, the baboons move out of the way at the last moment. We have seen wart hogs chasing each other, and a running rhinoceros go right through a troop, with the baboons merely stepping out of the way. We have seen male impalas fighting while baboons fed beside them. Once we saw a baboon chase a giraffe, but it seemed to be more in play than aggression.

Only rarely did we see baboons engage in hostilities against other species. On one occasion, however, we saw a baboon kill a small vervet monkey and eat it. The vervets frequented the same water holes as the baboons and usually they moved near them or even among them without incident. But one troop of baboons we observed at Victoria Falls pursued vervets on sight and attempted, without success, to keep

BABOON EATS A POTATO tossed to him by a member of the authors' party. Baboons are primarily herbivores but occasionally they will eat birds' eggs and even other animals.

INFANT BABOON rides on its mother's back through a park outside Nairobi. A newborn infant first travels by clinging to its mother's chest, but soon learns to ride pickaback.

BABOONS AND THREE OTHER SPECIES gather near a water hole (*out of picture to right*). Water holes and the relatively lush vegetation that surrounds them are common meeting places for a wide variety of herbivores. In this scene of the open savanna of

them out of certain fruit trees. The vervets easily escaped in the small branches of the trees.

The baboons' food is almost entirely vegetable, although they do eat meat on rare occasions. We saw dominant males kill and eat two newborn Thomson's gazelles. Baboons are said to be fond of fledglings and birds' eggs and have even been reported digging up crocodile eggs. They also eat insects. But their diet consists principally of grass, fruit, buds and plant shoots of many kinds; in the Nairobi area alone they consume more than 50 species of plant.

For baboons, as for many herbivores, association with other species on the range often provides mutual protection. In open country their closest relations are with impalas, while in forest areas the bushbucks play a similar role. The ungulates have a keen sense of smell, and baboons have keen eyesight. Baboons are visually alert, constantly looking in all directions as they feed. If they see predators, they utter warning barks that alert not only the other baboons but also any other animals that may be in the vicinity. Similarly, a warning bark by a bushbuck or an impala will put a baboon troop to flight. A mixed herd of impalas and baboons is almost impossible to take by surprise.

Impalas are a favorite prey of cheetahs. Yet once we saw impalas, grazing in the company of baboons, make no effort to escape from a trio of approaching cheetahs. The impalas just watched as an adult male baboon stepped toward the cheetahs, uttered a cry of defiance and sent them trotting away.

The interdependence of the different species is plainly evident at a water hole, particularly where the bush is thick and visibility poor. If giraffes are drinking, zebras will run to the water. But the first animals to arrive at the water hole approach with extreme caution. In the Wankie reserve, where we also observed baboons, there are large water holes

the Amboseli reserve there are baboons in the foreground and middle distance. An impala moves across the foreground just left of center. A number of zebras are present; groups of gnus graze together at right center and move off toward the water hole (right).

surrounded by wide areas of open sand between the water and the bushes. The baboons approached the water with great care, often resting and playing for some time in the bushes before making a hurried trip for a drink. Clearly, many animals know each other's behavior and alarm signals.

A baboon troop finds its ultimate safety, however, in the trees. It is no exaggeration to say that trees limit the distribution of baboons as much as the availability of food and water. We observed an area by a marsh in Amboseli where there was water and plenty of food. But there were lions and no trees and so there were no baboons. Only a quarter of a mile away, where lions were seen even more frequently, there were trees. Here baboons were numerous; three large troops frequented the area.

At night, when the carnivores and snakes are most active, baboons sleep high up in big trees. This is one of the baboon's primary behavioral adaptations. Diurnal living, together with an arboreal refuge at night, is an extremely effective way for them to avoid danger. The callused areas on a baboon's haunches allow it to sleep sitting up, even on small branches; a large troop can thus find sleeping places in a few trees. It is known that Colobus monkeys have a cycle of sleeping and waking throughout the night; baboons probably have a similar pattern. In any case, baboons are terrified of the dark. They arrive at the trees before night falls and stay in the branches until it is fully light. Fear of the dark, fear of falling and fear of snakes seem to be basic parts of the primate heritage.

Whether by day or night, individual baboons do not wander away from the troop, even for a few hours. The importance of the troop in ensuring the survival of its members is dramatized by the fate of those that are badly injured or too sick to keep up with their fellows. Each day the troop travels on a circuit

LIONESS LEAPS AT A THORN TREE into which a group of baboons has fled for safety. Lions appear to be among the few animals that successfully prey on baboons. The car in the background drove up as the authors' party was observing the scene.

of two to four miles; it moves from the sleeping trees to a feeding area, feeds, rests and moves again. The pace is not rapid, but the troop does not wait for sick or injured members. A baby baboon rides its mother, but all other members of the troop must keep up on their own. Once an animal is separated from the troop the chances of death are high. Sickness and injuries severe enough to be easily seen are frequent. For example, we saw a baboon with a broken forearm. The hand swung uselessly, and blood showed that the injury was recent. This baboon was gone the next morning and was not seen again. A sickness was widespread in the Amboseli troops, and we saw individuals dragging themselves along, making tremendous efforts to stay with the troop but falling behind. Some of these may have rejoined their troops; we are sure that at least five did not. One sick little juvenile lagged for four days and then apparently recovered. In the somewhat less natural setting of Nairobi park we saw some baboons that

had lost a leg. So even severe injury does not mean inevitable death. Nonetheless, it must greatly decrease the chance of survival.

Thus, viewed from the outside, the troop is seen to be an effective way of life and one that is essential to the survival of its individual members. What do the internal events of troop life reveal about the drives and motivations that cause individual baboons to "seek safety in numbers"? One of the best ways to approach an understanding of the behavior patterns within the troop is to watch the baboons when they are resting and feeding quietly.

Most of the troop will be gathered in small groups, grooming each other's fur or simply sitting. A typical group will contain two females with their young offspring, or an adult male with one or more females and juveniles grooming him. Many of these groups tend to persist, with the same animals that have been grooming each other walking together when the troop moves. The nucleus of

such a "grooming cluster" is most often a dominant male or a mother with a very young infant. The most powerful males are highly attractive to the other troop members and are actively sought by them. In marked contrast, the males in many ungulate species, such as impalas, must constantly herd the members of their group together. But baboon males have no need to force the other troop members to stay with them. On the contrary, their presence alone ensures that the troop will stay with them at all times.

Young infants are equally important in the formation of grooming clusters. The newborn infant is the center of social attraction. The most dominant adult males sit by the mother and walk close beside her. When the troop is resting, adult females and juveniles come to the mother, groom her and attempt to groom the infant. Other members of the troop are drawn toward the center thus formed, both by the presence of the pro-

BABOONS AND IMPALAS cluster together around a water hole. The two species form a mutual alarm system. The baboons have keen eyesight and the impalas a good sense of smell. Between them they quickly sense the presence of predators and take flight.

tective adult males and by their intense interest in the young infants.

In addition, many baboons, especially adult females, form preference pairs, and juvenile baboons come together in play groups that persist for several years. The general desire to stay in the troop is strengthened by these "friendships," which express themselves in the daily pattern of troop activity.

Our field observations, which so strongly suggest a high social motivation, are backed up by controlled experiment in the laboratory. Robert A. Butler of Walter Reed Army Hospital has shown that an isolated monkey will work hard when the only reward for his labor is the sight of another monkey [see "Curiosity in Monkeys," by Robert A. Butler; SCIENTIFIC AMERICAN Offprint 426]. In the troop this social drive is expressed in strong individual prefer-

ences, by "friendship," by interest in the infant members of the troop and by the attraction of the dominant males. Field studies show the adaptive value of these social ties. Solitary animals are far more likely to be killed, and over the generations natural selection must have favored all those factors which make learning to be sociable easy.

The learning that brings the individual baboon into full identity and participation in the baboon social system begins with the mother-child relationship. The newborn baboon rides by clinging to the hair on its mother's chest. The mother may scoop the infant on with her hand, but the infant must cling to its mother, even when she runs, from the day it is born. There is no time for this behavior to be learned. Harry F. Harlow of the University of Wisconsin has shown that an infant monkey will auto-

matically cling to an object and much prefers objects with texture more like that of a real mother [see the article "Love in Infant Monkeys," by Harry F. Harlow; SCIENTIFIC AMERICAN Offprint 429]. Experimental studies demonstrate this clinging reflex; field observations show why it is so important.

In the beginning the baboon mother and infant are in contact 24 hours a day. The attractiveness of the young infant, moreover, assures that he and his mother will always be surrounded by attentive troop members. Experiments show that an isolated infant brought up in a laboratory does not develop normal social patterns. Beyond the first reflexive clinging, the development of social behavior requires learning. Behavior characteristic of the species depends therefore both on the baboon's biology and on the social situations that are present in the troop.

As the infant matures it learns to ride on its mother's back, first clinging and then sitting upright. It begins to eat solid foods and to leave the mother for longer and longer periods to play with other infants. Eventually it plays with the other juveniles many hours a day, and its orientation shifts from the mother to this play group. It is in these play groups that the skills and behavior patterns of adult life are learned and practiced. Adult gestures, such as mounting, are frequent, but most play is a mixture of chasing, tail-pulling and mock fighting. If a juvenile is hurt and cries out, adults come running and stop the play. The presence of an adult male prevents small juveniles from being hurt. In the protected atmosphere of the play group the social bonds of the infant are widely extended.

Grooming, a significant biological function in itself, helps greatly to establish social bonds. The mother begins grooming her infant the day it is born, and the infant will be occupied with grooming for several hours a day for the rest of its life. All the older baboons do a certain amount of grooming, but it is the adult females who do most. They groom the infants, juveniles, adult males and other females. The baboons go to each other and "present" themselves for grooming. The grooming animal picks through the hair, parting it with its hands, removing dirt and para-sites, usually by nibbling. Grooming is most often reciprocal, with one animal doing it for a while and then presenting itself for grooming. The animal being groomed relaxes, closes its eyes and gives every indication of complete pleasure. In addition to being pleasurable, grooming serves the important function of keeping the fur clean. Ticks are common in this area and can be seen on many animals such as dogs and lions; a baboon's skin, however, is free of them. Seen in this light, the enormous amount of time baboons spend in grooming each other is understandable. Grooming is pleasurable to the individual, it is the most important expression of close social bonds and it is biologically adaptive.

The adults in a troop are arranged in a dominance hierarchy, explicitly revealed in their relations with other members of the troop. The most dominant males will be more frequently groomed and they occupy feeding and resting positions of their choice. When a dominant animal approaches a subordinate one, the lesser animal moves out of the way. The observer can determine the order of dominance simply by watching the reactions of the baboons as they move past each other. In the tamer troops these observations can be tested by feeding. If food is tossed between two baboons, the more dominant one will take it, whereas the other may not even look at it directly.

The status of a baboon male in the dominance hierarchy depends not only on his physical condition and fighting ability but also on his relationships with other males. Some adult males in every large troop stay together much of the time, and if one of them is threatened, the others are likely to back him up. A group of such males outranks any individual, even though another male outside the group might be able to defeat any member of it separately. The hierarchy has considerable stability and this is due in large part to its dependence on clusters of males rather than the fighting ability of individuals. In troops where the rank order is clearly defined, fighting is rare. We observed frequent bickering or severe fighting in only about 15 per cent of the troops. The usual effect of the hierarchy, once relations among the males are settled, is to decrease disruptions in the troop. The dominant animals, the males in particular, will not let others fight. When bickering breaks out, they usually run to the scene and stop it. Dominant males thus protect the weaker animals against harm from inside as well as outside. Females and juveniles come to the males to groom them or just to sit beside them. So although dominance depends ultimately on force, it leads to peace, order and popularity.

	ECOLOGY			ECONOMIC SYSTEM	
	GROUP SIZE, DENSITY AND RANGE	HOME BASE	POPULATION STRUCTURE	FOOD HABITS	ECONOMIC DEPENDENCE
	GROUPS OF 50–60 COMMON BUT VARY WIDELY. ONE INDIVIDUAL PER 5–10 SQUARE MILES. RANGE 200–600 SQUARE MILES. TERRITORIAL RIGHTS; DEFEND BOUNDARIES AGAINST STRANGERS.	OCCUPY IMPROVED SITES FOR VARIABLE TIMES WHERE SICK ARE CARED FOR AND STORES KEPT.	TRIBAL ORGANIZATION OF LOCAL, EXOGAMOUS GROUPS.	OMNIVOROUS. FOOD SHARING. MEN SPECIALIZE IN HUNTING, WOMEN AND CHILDREN IN GATHERING.	INFANTS ARE DEPENDENT ON ADULTS FOR MANY YEARS. MATURITY OF MALE DELAYED BIOLOGICALLY AND CULTURALLY. HUNTING, STORAGE AND SHARING OF FOOD.
	10–200 IN GROUP. 10 INDIVIDUALS PER SQUARE MILE. RANGE 3–6 SQUARE MILES; NO TERRITORIAL DEFENSE.	NONE: SICK AND INJURED MUST KEEP UP WITH TROOP.	SMALL, INBREEDING GROUPS.	ALMOST ENTIRELY VEGETARIAN. NO FOOD SHARING, NO DIVISION OF LABOR.	INFANT ECONOMICALLY INDEPENDENT AFTER WEANING. FULL MATURITY BIOLOGICALLY DELAYED. NO HUNTING, STORAGE OR SHARING OF FOOD.

APES AND MEN are contrasted in this chart, which indicates that although apes often seem remarkably "human," there are fundamental differences in behavior. Baboon characteristics, which may be taken as representative of ape and monkey behavior in

Much has been written about the importance of sex in uniting the troop, it has been said, for example, that "the powerful social magnet of sex was the major impetus to subhuman primate sociability" [see "The Origin of Society," by Marshall D. Sahlins; SCIENTIFIC AMERICAN Offprint 602]. Our observations lead us to assign to sexuality a much lesser, and even at times a contrary, role. The sexual behavior of baboons depends on the biological cycle of the female. She is receptive for approximately one week out of every month, when she is in estrus. When first receptive, she leaves her infant and her friendship group and goes to the males, mating first with the subordinate males and older juveniles. Later in the period of receptivity she goes to the dominant males and "presents." If a male is not interested, the female is likely to groom him and then present again. Near the end of estrus the dominant males become very interested, and the female and a male form a consort pair. They may stay together for as little as an hour or for as long as several days. Estrus disrupts all other social relationships, and consort pairs usually move to the edge of the troop. It is at this time that fighting may take place, if the dominance order is not clearly established among the males. Normally there is no fighting over females, and a male, no matter how dominant, does not monopolize a female

for long. No male is ever associated with more than one estrus female; there is nothing resembling a family or a harem among baboons.

Much the same seems to be true of other species of monkey. Sexual behavior appears to contribute little to the cohesion of the troop. Some monkeys have breeding seasons, with all mating taking place within less than half the year. But even in these species the troop continues its normal existence during the months when there is no mating. It must be remembered that among baboons a female is not sexually receptive for most of her life. She is juvenile, pregnant or lactating; estrus is a rare event in her life. Yet she does not leave the troop even for a few minutes. In baboon troops, particularly small ones, many months may pass when no female member comes into estrus; yet no animals leave the troop, and the highly structured relationships within it continue without disorganization.

The sociableness of baboons is expressed in a wide variety of behavior patterns that reinforce each other and give the troop cohesion. As the infant matures the nature of the social bonds changes continually, but the bonds are always strong. The ties between mother and infant, between a juvenile and its peers in a play group, and between a mother and an adult male are quite different from one another. Similarly, the

bond between two females in a friendship group, between the male and female in a consort pair or among the members of a cluster of males in the dominance hierarchy is based on diverse biological and behavioral factors, which offer a rich field for experimental investigation.

In addition, the troop shares a considerable social tradition. Each troop has its own range and a secure familiarity with the food and water sources, escape routes, safe refuges and sleeping places inside it. The counterpart of the intensely social life within the troop is the coordination of the activities of all the troop's members throughout their lives. Seen against the background of evolution, it is clear that in the long run only the social baboons have survived.

When comparing the social behavior of baboons with that of man, there is little to be gained from laboring the obvious differences between modern civilization and the society of baboons. The comparison must be drawn against the fundamental social behavior patterns that lie behind the vast variety of human ways of life. For this purpose we have charted the salient features of baboon life in a native habitat alongside those of human life in preagricultural society [see chart below]. Cursory inspection shows that the differences are more numerous and significant than are the

SOCIAL SYSTEM					COMMUNICATION
ORGANIZATION	SOCIAL CONTROL	SEXUAL BEHAVIOR	MOTHER-CHILD RELATIONSHIP	PLAY	
BANDS ARE DEPENDENT ON AND AFFILIATED WITH ONE ANOTHER IN A SEMIOPEN SYSTEM. SUBGROUPS BASED ON KINSHIP.	BASED ON CUSTOM.	FEMALE CONTINUOUSLY RECEPTIVE. FAMILY BASED ON PROLONGED MALE-FEMALE RELATIONSHIP AND INCEST TABOOS.	PROLONGED; INFANT HELPLESS AND ENTIRELY DEPENDENT ON ADULTS.	INTERPERSONAL BUT ALSO CONSIDERABLE USE OF INANIMATE OBJECTS.	LINGUISTIC COMMUNITY. LANGUAGE CRUCIAL IN THE EVOLUTION OF RELIGION, ART, TECHNOLOGY AND THE CO-OPERATION OF MANY INDIVIDUALS.
TROOP SELF-SUFFICIENT, CLOSED TO OUTSIDERS. TEMPORARY SUBGROUPS ARE FORMED BASED ON AGE AND INDIVIDUAL PREFERENCES.	BASED ON PHYSICAL DOMINANCE.	FEMALE ESTRUS. MULTIPLE MATES. NO PROLONGED MALE-FEMALE RELATIONSHIP.	INTENSE BUT BRIEF; INFANT WELL DEVELOPED AND IN PARTIAL CONTROL.	MAINLY INTERPERSONAL AND EXPLORATORY.	SPECIES-SPECIFIC, LARGELY GESTURAL AND CONCERNED WITH IMMEDIATE SITUATIONS.

general, are based on laboratory and field studies; human characteristics are what is known of preagricultural Homo sapiens. The chart suggests that there was a considerable gap between primate behavior and the behavior of the most primitive men known.

BABOONS AND ELEPHANTS have a relationship that is neutral rather than co-operative, as in the case of baboons and impalas. If an elephant or another large herbivore such as a rhinoceros moves through a troop, the baboons merely step out of the way.

similarities.

The size of the local group is the only category in which there is not a major contrast. The degree to which these contrasts are helpful in understanding the evolution of human behavior depends, of course, on the degree to which baboon behavior is characteristic of monkeys and apes in general and therefore probably characteristic of the apes that evolved into men. Different kinds of monkey do behave differently, and many more field studies will have to be made before the precise degree of difference can be understood.

For example, many arboreal monkeys have a much smaller geographical range than baboons do. In fact, there are important differences between the size and type of range for many monkey species. But there is no suggestion that a troop of any species of monkey or ape occupies the hundreds of square miles ordinarily occupied by preagricultural human societies. Some kinds of monkey may resent intruders in their range more than baboons do, but there is no evidence that any species fights for complete control of a territory. Baboons are certainly less vocal than some other monkeys, but no nonhuman primate has even the most rudimentary language. We believe that the fundamental contrasts in our chart would hold for the vast majority of monkeys and apes as compared with the ancestors of man. Further study of primate behavior will sharpen these contrasts and define more clearly the gap that had to be traversed from ape to human behavior. But already we can see that man is as unique in his sharing, co-operation and play patterns as he is in his locomotion, brain and language.

The basis for most of these differences may lie in hunting. Certainly the hunting of large animals must have involved co-operation among the hunters and sharing of the food within the tribe. Similarly, hunting requires an enormous extension of range and the protection of a hunting territory. If this speculation proves to be correct, much of the evolution of human behavior can be reconstructed, because the men of 500,000 years ago were skilled hunters. In locations such as Choukoutien in China and Olduvai Gorge in Africa there is evidence of both the hunters and their campsites [see "Olduvai Gorge," by L. S. B. Leakey; SCIENTIFIC AMERICAN, January, 1954]. We are confident that the study of the living primates, together with the archaeological record, will eventually make possible a much richer understanding of the evolution of human behavior.

GENERAL BIBLIOGRAPHY

1. General Introduction to Animal Social Behavior

THE SOCIAL LIFE OF INSECTS: THEIR ORIGIN AND EVOLUTION. W. M. Wheeler. Harcourt, Brace, 1928.

THE SOCIAL LIFE OF ANIMALS. W. C. Allee. W. W. Norton, 1938.

THE INSECT SOCIETIES. E. O. Wilson. The Belknap Press of Harvard University Press, 1971.

SELECTED WRITINGS OF T. C. SCHNEIRLA. Edited by L. R. Aronson, E. Tobach, J. S. Rosenblatt, and D. S. Lehrman. W. H. Freeman, 1972.

BIOLOGICAL BASES OF HUMAN SOCIAL BEHAVIOR. R. A. Hinde. McGraw-Hill, 1974.

HOW ANIMALS COMMUNICATE. T. A. Sebeok. Indiana University Press, 1977.

2. Introduction to Sociobiological Theory

SOCIOBIOLOGY: THE NEW SYNTHESIS. E. O. Wilson. The Belknap Press of Harvard University Press, 1975.

READINGS IN SOCIOBIOLOGY. Edited by T. H. Clutton-Brock and P. H. Harvey. W. H. Freeman, 1978.

THE SOCIOBIOLOGY DEBATE: READINGS ON THE ETHICAL AND SCIENTIFIC ISSUES CONCERNING SOCIOBIOLOGY. Edited by A. L. Caplan. Harper & Row, 1978.

BIBLIOGRAPHIES

1. Animal Communication

MECHANISMS OF ANIMAL BEHAVIOR. P. R. Marler and W. J. Hamilton III. John Wiley & Sons, Inc., 1966.

ANIMAL COMMUNICATION: TECHNIQUES OF STUDY AND RESULTS OF RESEARCH. Edited by T. A. Sebeok. Indiana University Press, 1968.

SEX PHEROMONE SPECIFICITY: TAXONOMIC AND EVOLUTIONARY ASPECTS IN LEPIDOPTERA. Wendell L. Roelofs and Andre Comeau in Science, Vol. 165, No. 3891, pages 398–400; July 25, 1969.

THE INSECT SOCIETIES. Edward O. Wilson. The Belknap Press of Harvard University Press, 1971.

LANGUAGE IN CHIMPANZEE? David Premack in Science, Vol. 172, No. 3985, pages 802–822; May 21, 1971.

NON-VERBAL COMMUNICATION. Edited by R. A. Hinde. Cambridge University Press, 1972.

2. The Evolution of Behavior

THE GENETICAL EVOLUTION OF SOCIAL BEHAVIOUR. W. D. Hamilton in Journal of Theoretical Biology, Vol. 7, pages 1–52; 1964.

THE SELFISH GENE. Richard Dawkins. Oxford University Press, 1976.

3. Social Spiders

THE INSECT SOCIETIES. Edward O. Wilson. The Belknap Press of Harvard University Press, 1971.

EVOLUTION OF SOCIAL BEHAVIOR IN SPIDERS (ARANEAE; ERESITAE AND THERIDIIDAE). Ernst J. Kullmann in American Zoologist, Vol. 12, No. 3, pages 419–426; August, 1972.

THE SHEET WEB AS A TRANSDUCER, MODIFYING VIBRATION SIGNALS IN SOCIAL SPIDER COLONIES OF MALLOS GREGALIS. J. Wesley Burgess in Neuroscience Abstracts, Society for Neuroscience Fifth Annual Meeting; 1975.

4. The Social Behavior of Army Ants

BEHAVIORAL STUDIES OF ARMY ANTS. Carl W. Rettenmeyer in University of Kansas Science Bulletin, Vol. 44, pages 281–465; September, 1963.

ARMY ANTS: A STUDY IN SOCIAL ORGANIZATION. T. C. Schneirla. Edited by Howard Topoff. W. H. Freeman, 1971.

POLYMORPHISM IN ARMY ANTS RELATED TO DIVISION OF LABOR AND COLONY CYCLIC BEHAVIOR. Howard Topoff in The American Naturalist, Vol. 105, No. 946, pages 529–548; November–December, 1971.

5. Weaver Ants

THE INSECT SOCIETIES. Edward O. Wilson. The Belknap Press of Harvard University Press, 1971.

WEAVER ANTS: SOCIAL ESTABLISHMENT AND MAINTENANCE OF TERRITORY. Berthold K. Hölldobler and Edward O. Wilson in Science, Vol. 195, No. 4281, pages 900–902; March 4, 1977.

COLONY-SPECIFIC TERRITORIAL PHEROMONE IN THE AFRICAN WEAVER ANT OECOPHYLLA LONGINODA (LATREILLE). Berthold K. Hölldobler and Edward O. Wilson in Proceedings of the National Academy of Sciences, Vol. 74, No. 5, pages 2072–2075; May, 1977.

6. The Schooling of Fishes

STUDIES ON THE STRUCTURE OF THE FISH SCHOOL. C. M. Breder, Jr., in Bulletin of the American Museum of Natural History, Vol. 98, Art. 1, pages 1–27; 1951.

STUDIES ON SOCIAL GROUPINGS IN FISHES. C. M. Breder, Jr., in Bulletin of the American Museum of Natural History, Vol. 117, Art. 6, pages 393–481; 1959.

THE DEVELOPMENT OF SCHOOLING BEHAVIOR IN FISHES. Evelyn Shaw in *Physiological Zoology*, Vol. 33, No. 2, pages 79–86; 1960.

THE DEVELOPMENT OF SCHOOLING IN FISHES II. Evelyn Shaw in *Physiological Zoology*, Vol. 34, No. 4, pages 263–272; 1961.

7. "Imprinting" in a Natural Laboratory

"IMPRINTING" IN ANIMALS. Eckhard H. Hess in *Scientific American*, Vol. 198, No. 3, pages 81–90; March, 1958.

IMPRINTING IN BIRDS. Eckhard H. Hess in *Science*, Vol. 146, No. 3648, pages 1128–1139; November 27, 1964.

INNATE FACTORS IN IMPRINTING. Eckhard H. Hess and Dorle B. Hess in *Psychonomic Science*, Vol. 14, No. 3, pages 129–130; February 10, 1969.

DEVELOPMENT OF SPECIES IDENTIFICATION IN BIRDS: AN INQUIRY INTO THE PRENATAL DETERMINANTS OF PERCEPTION. Gilbert Gottlieb. University of Chicago Press, 1971.

NATURAL HISTORY OF IMPRINTING. Eckhard H. Hess in *Integrative Events in Life Processes: Annals of the New York Academy of Sciences*, Vol. 193, pages 124–136; 1972.

8. The Social Ecology of Coyotes

THE WOLF: THE ECOLOGY AND BEHAVIOR OF AN ENDANGERED SPECIES. L. David Mech. American Museum of Natural History, Natural History Press, 1970.

FEEDING AND SOCIAL BEHAVIOR OF THE STRIPED HYENA (*HYAENA VULGARIS*). H. Kruuk in *East African Wildlife Journal*, Vol. 14, pages 91–111; 1976.

BEHAVIORAL ECOLOGY. Edited by J. R. Krebs and N. B. Davies. Sinauer Associates, Inc., 1978.

THE CLEVER COYOTE. Stanley P. Young and Hartley H. T. Jackson. University of Nebraska Press, 1978.

COYOTES: BIOLOGY, BEHAVIOR, AND MANAGEMENT. Edited by Marc Bekoff. Academic Press, 1978.

THE RELATIONSHIP BETWEEN FOOD COMPETITION AND FORAGING SIZE GROUP IN SOME LARGER CARNIVORES. J. Lamprecht in *Zeitschrift für Tierpsychologie*, Vol. 46, pages 337–343; 1978.

9. Dolphins

MAMMALS OF THE SEA: BIOLOGY AND MEDICINE. Edited by Sam H. Ridgway. Charles C. Thomas, Publisher, 1972.

OCCURRENCE AND GROUP ORGANIZATION OF ATLANTIC BOTTLENOSE PORPOISES (*TURSIOPS TRUNCATUS*) IN AN ARGENTINE BAY. Bernd Würsig in *The Biological Bulletin*, Vol. 154, No. 2, pages 348–359; 1978.

10. The Social Life of Baboons

BEHAVIOR AND EVOLUTION. Edited by Anne Roe and George Gaylord Simpson. Yale University Press, 1958.

A STUDY OF BEHAVIOUR OF THE CHACMA BABOON, *PAPIO URSINUS*. Niels Bolwig in *Behaviour*, Vol. 14, No. 1–2, pages 136–163; 1959.

GLOSSARY

Allele   An alternative form of the same gene locus.

Altruism   Behavior disadvantageous to the individual, but beneficial to other members of the species.

Brood   A collective term for the immature stages (eggs, larvae, pupae) of a holometabolous insect.

Callow   An insect that has recently emerged from the pupal stage of development; its exoskeleton is relatively soft and unpigmented.

Clutch   All of the eggs laid and incubated by a single bird at one time.

Communal   In insects, living as a colony consisting of females of a single generation, each making, provisioning, and laying eggs in her own cell.

Consanguinity   The degree of genetic relatedness of individuals who have a common ancestor.

Diploid   Having a maternal and paternal set of chromosomes, resulting from an egg being fertilized.

Displacement activity   The performance of a behavioral act, often in a conflict situation, that appears not to be relevant to the prevailing situation.

Display   A pattern of behavior that has been modified in the course of evolution to function in intraspecific communication.

Dominance order   A behaviorally maintained hierarchy existing within animal groups, typically involving access to mates, food, and other resources.

Echolocation   A sonarlike system used by some species of bats, birds, and porpoises for both orientation and communication.

Emigration   The movement of a colony of ants from one nesting site to another.

Estrous cycle   In female mammals, rhythmic variations in reproductive physiology and behavior, due primarily to the periodic secretion of gonadotrophic hormones.

Eusocial   In insects, a degree of social behavior characterized by cooperative care of the young, division of labor, and an overlap of at least two generations.

Haplodiploidy   A mode of sex determination, found in the insect order Hymenoptera, in which males develop from unfertilized eggs and females from fertilized eggs.

Haploid   Having only a single set of chromosomes, usually resulting from the process of meiosis.

Holometabolous   Undergoing a complete metamorphosis during development, with distinct larval, pupal, and adult stages.

Home range   The area in which an animal might be found at any given time.

Imprinting   The strong attachment formed by many young animals for their parents or other moving objects shortly after birth.

Inclusive fitness   The sum of an individual's own reproductive fitness, plus all influences on fitness in its relatives.

Kin selection   The selection for activities that lower an individual's own reproduction, but increase the fitness of relatives.

Larva   An immature stage of insect development, in which the organism is radically different in form from the adult organism.

Locus   Position of a gene on a chromosome.

Meiosis   A series of two modified mitoses, generating haploid gametes from a diploid precursor cell.

Mitosis   The process of cell division in which duplication and assortment of the chromosomes insures that all resulting cells have the same genetic information.

Pheromone   A chemical secreted by one individual that can affect the physiology and behavior of another member of the species.

Physogastric   In social insects, an enlarged condition of the queen's abdomen during the period of egg laying.

Pipping   The process by which a hatching bird breaks through the shell of an egg.

Polymorphism   In a colony of social insects, the existence of individuals that differ in both size and structure.

Pupa   The relatively inactive stage of holometabolous insects during which development into the adult form is completed.

Quasi-social   Living in a colony in which a group of females of the same generation cooperatively construct and provision nest cells.

Queen substance   The group of pheromones by which a queen honeybee regulates the reproductive activities of the worker bees.

Recruitment   In social insects, a process of communication in which nestmates are rapidly directed to food sources, new nests, or areas of colony disturbance.

Reproductive isolating mechanism   A property of individuals that prevents successful interbreeding with members of different populations.

Semisocial   Living as a colony consisting of a group of females of the same generation showing caste and division of labor.

Subsocial   Living as a colony consisting of an adult female and her immature offspring, which are progressively fed.

Territory   A fixed area from which rivals are excluded by active efforts of the resident animal.

Waggle dance   A pattern of behavior in honeybees that functions to communicate the location of feeding sites and new nests.

INDEX